# Perspectives in Cognitive Neuroscience

Stephen M. Kosslyn, *General Editor*

# Attentional Processing

—

*The Brain's Art of Mindfulness*

David LaBerge

Harvard University Press

Cambridge, Massachusetts
London, England
1995

Copyright © 1995 by the President and Fellows of Harvard College
All rights reserved
*Printed in the United States of America*

This book is printed on acid-free paper, and its binding materials
have been chosen for strength and durability.

Library of Congress Cataloging-in-Publication Data

LaBerge, David.
    Attentional processing : the brain's art of mindfulness / David
LaBerge.
        p.  cm. — (Perspectives in cognitive neuroscience)
    Includes bibliographical references and index.
    ISBN 0-674-05268-4
    1. Attention.   2. Neuropsychology.   I. Title.   II. Series.
QP405.L33 1995
    153.7'33—dc20        94-38071

*To Jan, my life partner*

# Contents

# Acknowledgments

——————

When Steve Kosslyn invited me to write on attention for the Perspectives in Cognitive Neuroscience series with Harvard University Press, I had already begun to outline the kind of book appropriate for the series: a synthesis of what the diverse fields of psychology, computer science, and neurobiology were saying about attentional processing. This book could not have been written without the support and encouragement of many people.

Over numerous years the University of Minnesota and the University of California, Irvine, generously provided the means to carry out my research on attention. Research support has also been provided by grants from the National Science Foundation, Joseph L. Young, Program Director, and from the Office of Naval Research, Harold Hawkins, Program Director. Much of the writing was done on sabbatical leave in the Berkshires of Massachusetts, and the library at nearby Simon's Rock College provided many needed references. For these library privileges I thank the Vice President of the College, Bernard Rodgers, the Head of the Library, Joan Goodkind, and the Reference Librarian, Russell Miller.

The research from my laboratory reported here was done in collaboration with my graduate students Blynn Bunney, Robert Carlson, Marc Carter, and John Williams, and in particular Vincent Brown. The work of our research group benefited from the contributions of Professor Alan Hartley of Scripps College.

I am greatly indebted to several colleagues for stimulating discussions of ideas that have found their way into this book: Aric Agmon, Monte Buchsbaum, Mary Baird Carlsen, Peter Fox, Michael Gazzaniga, Michael Goldberg, Patricia Goldman-Rakic, Richard Haier, Steven Hillyard, Herbert Killackey, Christof Koch, John Lawry, Mario Liotti, Duncan Luce, David McCormick, Steven Petersen,

Michael Posner, Steven Potkin, Murray Sherman, James Swanson, Sharon Wigal, Tim Wigal, Marty Woldorff and Joseph Wu. I especially thank James Carlsen for many valuable discussions of attention and expectancy in music, John Compton for tenaciously emphasizing the importance of attention's experiential aspects, William K. Estes for his encouragement and inspiring example of clarity in thinking, Ted Jones for generously providing his expertise on matters of the thalamus, and Stephen M. Kosslyn, general editor of the Perspectives in Cognitive Neuroscience series, for his thorough and valuable critique of the manuscript.

I am very grateful to my editors at Harvard University Press: Kate Schmit, who clarified obscure sentences, smoothed out lumpy syntax, and cheerfully moved me through the final stages of preparation, and Angela Von der Lippe, who guided and encouraged my efforts throughout the writing of the book.

Finally, I thank my wife, Janice Lawry, who found the words for the book's subtitle and provided inspiration, support, and patience over years of writing.

# 1

# Introduction

Our mental experience can be described as a stream of activities that includes perceptions, memories, feelings, intentions, images, beliefs, and desires. Several of these activities may be going on at the same time, as when we read a quotation in the newspaper, recognize it as a line from Shakespeare, kindle the meanings of the words as they are read, and experience the feelings that they arouse in us. On many occasions, however, one particular mental activity seems to "fill the mind," if only for a brief moment of time. Tasting a good cup of coffee in the morning, remembering a discussion with a friend, considering that the economy is on an upswing and wishing that it would continue—these experiences may each take a turn at occupying one's mind. Imagine an individual's ongoing experience as a river containing several currents of distinctive colors, and giving attention to one current expands its width as it flows along; if we could observe the river in a thin cross-section at this point of "mindfulness," it would seem to have only the one color.

This informal impression of the enhancing or magnifying property of attention, with which we are all familiar from private experience, appears to be borne out when the "mind," or mental processing, is observed by the objective methods of the psychological and neurobiological laboratories.

The term *mind* points to a variety of functions of the brain—thinking, feeling, intending, perceiving, judging, and so on—whereas the term *mindfulness* or *attention* points to the characteristic way in which any of these functions can move to center stage (or can move other

1

functions off stage) at any given moment. But the emphasis conferred on a particular mental activity as we attend to it or become
mindful of it can vary continuously from a high intensity, when the
mind seems completely occupied by that process, to a zero intensity,
when that process is carried out automatically. When we first learn
to read, we give all our attention to the perception of individual
words, but with practice the identification of words requires less attentional emphasis and more attention can be directed to understanding what we are reading. This book is about how this "mental
emphasis" is useful to the individual, and about how the brain is
constructed to enable such emphasis to take place.

My approach here incorporates two assumptions: attention can be
expressed in many pathways of the brain, particularly in areas of the
cerebral cortex that serve specific cognitive activities such as reading
words or understanding sentences, and attentional activity in each
of these pathways can be intensified to varying degrees by signals
arising from a structure of the brain called the thalamus.

One function of the thalamus during waking appears to be the
enhancement or elevation of localized brain activities corresponding to mental functions. Anatomical details are postponed to later
sections of this book, but a few general observations about the relation of the thalamus to the cerebral cortex are appropriate at this
point. The thalamus is about the size of the end segment of the
little finger and is located in the interior region of the brain. From
this central position in the brain, the thalamus connects directly with
virtually all areas of the cortex, the thin, surface layer of the brain
that is rich in circuits that serve cognitive computations. The circuitry of the thalamus is apparently capable of enhancing activity in
a restricted set of cortical neurons. For example, it can elevate activity in sets of cells that code a green object while activities in neighboring cells that code blue objects remain at a lower level; or it may
enhance activity in cells that code the spatial location of a particular
letter of a printed word while the activities in cells that code neighboring letter locations remain at a lower level.

The elevation of activity in cortical neurons coding for a particular
item (whether it be the color green or a baby's cry from the nursery
or the perfect word to describe your mood) above the activity in
neurons coding for similar or neighboring items is assumed in this
book to express the way that attention selects one object or idea

when a group of objects or ideas is presented to our senses or to our cognitive consideration. As a general rule, attention is intensified when an item (an object or an idea) is accompanied by other items, particularly when the neighboring items are similar to the target item. We are familiar with experiencing an elevated concentration of attention when we reach for a pencil on a desk cluttered with other objects, and when we attempt to identify a face in a room crowded with other faces.

The process of attending to things, like breathing, is ever present in our waking lives. As noted over a hundred years ago by William James, "Every one knows what attention is. It is the taking possession by the mind, in clear and vivid form, of one out of what seem several simultaneously possible objects or trains of thought" (James, 1890, p. 403). Such familiarity, however, has not conferred a scientific understanding of how attention works. Among psychologists, attention has been characterized by a broad array of metaphors: a filter (Broadbent, 1958), effort (Kahneman, 1973), resources (Shaw and Shaw, 1978), a control process of short-term memory (Shiffrin and Schneider, 1977), orienting (Posner, 1980), conjoining object features (Treisman and Gelade, 1980), a spotlight that moves (Tsal, 1983), a gate (Reeves and Sperling, 1986), a zoom lens (Eriksen and St. James, 1986), and both a selective channel and a preparatory activity distribution (LaBerge and Brown, 1989). The existence of so many diverse views of attention among researchers in the field highlights the daunting mystery of how to describe the way attention works, regardless of how intimately acquainted we are with using it.

## A Multidisciplinary Approach

Some aspects of human biology, such as respiration and circulation, have yielded rather easily to our understanding, because access to the relevant anatomical structures and physiological functions is relatively direct. Others, such as perception, learning, decision-making, and attention, are much more difficult to investigate because the underlying anatomical structures are difficult to observe; neural structures, by which these activities operate, are very small, and physiological indicators of neural activity often produce ambiguous results, owing to the difficulty of isolating one cognitive activity from another.

Breathing is as familiar to us as attention, but what has given the researcher of breathing a massive head start over the researcher of attention is a clear conception of the way that breathing is expressed in bodily structures. Until recent times, researchers of a cognitive process such as attention or learning had no more than speculations of how the process might be expressed in brain structures, if they ventured to pose this question at all. We know that breathing is expressed as the movement of air into and out of the lungs (the goal being the exchange of $O_2$ and $CO_2$ gases), and circulation is expressed as the movement of fluid through the blood vessels of the body (the goal being the pickup and delivery of substances). With these concepts at hand, researchers could then move directly and easily to the discovery of the bellows-like mechanism responsible for respiration and the pump-like mechanism responsible for circulation, which led to an understanding of how these processes work.

Currently, an increasing number of researchers of cognitive activities are asking how these processes are expressed in the underlying biological hardware. In the field of learning, for example, it is now generally known that one kind of learning that occurs in the hippocampus (among other brain sites), long-term potentiation (LTP), is expressed as a change (reduction) in the number of discharges arriving at the input of a neuron needed to produce a discharge at the output of a neuron. ("Discharges" are transient electrical signals by which neurons communicate with other cells and send messages from one part of a cell to another.) A good deal of current research activity in the field of learning is now devoted to determining the mechanism that produces this expression of learning.

An important part of the approach to attention described in this book is the attempt to discover the way or ways in which attention is expressed in the brain. One way to move effectively toward this discovery is to ask what goals attention is expected to achieve for the system, or, put otherwise, what problems it is expected to solve to make the system more adaptable to its environment. Often answers to these questions are given in behavioral or cognitive terms and lead to the design of effective psychological tasks necessary for the testing of hypotheses concerning attention, or to the design of appropriate attention tasks to be carried out while physiological measures (such as positron emission tomography, or PET, scans) are taken. But it is also particularly helpful to answer these questions in terms of the types of computations needed to solve specific problems and then search the myriad kinds of activities expressed in brain pathways for these computation types.

Once the expression of attention in brain pathways is ascertained, the next step is to discover what mechanism or mechanisms produce the expression of attention. That is, the expression of attention in the brain is assumed to be produced by some mechanism that embodies or instantiates an appropriate algorithm whose output is the expression of attention in neural tissue. Having computational descriptions of attentional expression in hand can greatly reduce the number of candidate mechanisms.

But the cognitive-neuroscience story of attention, like the biological story of respiration and circulation, also requires a specification of what regulates or controls the mechanism that produces the immediate expression of the activity. Thus, the present approach to attention consists of four considerations: its *goals* (preferably stated in computational terms), its *expression(s)* in brain pathways, its *mechanism(s)*, and its sources of *control*.

Given these various parts of the problem, understanding the dismaying complexity of attention seems out of the reach of a single scientific discipline. Psychological research reveals "what is done" by attention as individuals respond adaptively to cognitive and behavioral task demands, but the history of psychology has not produced a single attentional metaphor or even a small set of interrelated metaphors on which most experts in the field agree. Philosophical considerations of "what it is like" experientially to be paying attention to something, which began this chapter, are usually framed in the context of the more general and probably more complex problem of consciousness, which has eluded our understanding for centuries.

On the other hand, neurobiological methods cannot be expected to reveal "how attention operates" if we have no idea of what sort of activity to look for, that is, what activity enables the system to solve the class of problems regarded as attentional. Furthermore, the computational approach of thinking up abstract algorithms (effective procedures) that describe "what is being computed" while paying attention could generate an endless list of possibilities if we were ignorant of the range of computations available to the brain and the kinds of behavioral problems attention solves for the individual.

In reaction to this state of affairs, many researchers of attention are beginning to combine some of the foregoing methodologies, in particular the methods of cognitive psychology, computer science, and neuroscience. Psychological tasks evoke attentional processing and can demonstrate its adaptive advantage in performing behavioral and cognitive tasks, while analyses of information input and output show what problems attention needs to solve to achieve these adaptive advantages. Together the psychological and computational analyses can guide the neuroscientist's search for brain areas and brain processes that implement attentional computations. Researchers who combine these methodologies in various ways have spawned the new interdisciplinary fields of inquiry of cognitive neuroscience, computational neuroscience, and cognitive science. This book attempts to combine cognitive, computational, and neuroscience research methods to build a testable theory of attention. At the same time, it uses the more experiential analyses of some philosophical approaches to ferret out important examples of attentional processing and to help maintain a synoptic view of attention in the face of highly detailed psychological and neurobiological data.

Combining methodologies and data from cognitive and biological sciences is, of course, not a new phenomenon among researchers. William James, whose work in cognitive psychology leaned heavily toward philosophy throughout his career, trained for and obtained the M.D. degree and laced his *Principles of Psychology* with references to "currents among the cells and fibers of his brain" (1890; see, e.g., vol. 2, p. 527). Since that time many psychological researchers have called themselves psychophysiologists.

Today, computational neuroscientists seek to explain how electrical and chemical signals are used in the brain to process information. Information is a well-defined notion for the computer: it is measured precisely in units of "bits" according to Shannon's logarithmic formula

$$I(s) = \log_2 \frac{1}{p(s)}$$

where $p(s)$ is the probability that a particular signal or event, $s$, will occur and $I(s)$ is the amount of information transmitted in bits when that signal or event occurs (Shannon and Weaver, 1949). The label of the information unit, *bit,* is a contraction of the words *binary digit,* and the expression of information units in binary form is made explicit in the equation by the use of the base-2 logarithm.

While the information in computers can be regarded as symbolic "representations" of messages or events, it is not clear that information in the nervous system "represents" something in the same sense (Freeman, 1975; see also Rumelhart and McClelland, 1986, on the notion of neural information processing as "subsymbolic"). Furthermore, psychological studies have shown that some human limitation in processing may be defined more appropriately in terms of "chunks" rather than bits (e.g., Miller, 1955); for example, humans can briefly remember about seven randomly presented letters, but if a very long sequence of letters is chunked into familiar words (e.g., appletablewindowtelevision), then humans in effect can remember a great many more letters. Computers, in contrast, have strict limits on the number of bits that can be stored in hardware and therefore cannot compress many bits into a single chunk in the way that humans do.

Although the nature of information processed by the nervous system may be fundamentally different from the "Shannon-information" processed by computers, both kinds of information are based on a form of signaling that, among other things, is independent of the energy level of the signal event. In general, a yes-or-no message conveys the same amount of information to a human whether it is whispered or shouted, and the same amount of information to a computer whether it is typed in with a weak or strong hand at the keyboard.

Information, however defined, is transferred within and between neurons by electrochemical events. The spatial and temporal patterns of these signal-like events are assumed to constitute computational processing just as the spatial and temporal coincidences of electrical pulses instantiate computations within an electrical computer.

The information that forms the ingredients of these computations is embodied in the sequence of signals conveyed along pathways within the computer and the brain. It turns out that the measure of Shannon-information in a sequence of signals in a specific computer pathway can almost always be precisely related to the general input-output functions of the computer (one misspelled word can prevent a program from running), while a Shannon-information measure of the sequence of discharges in a specific brain pathway rarely has so direct a relationship with human cognitive or behavioral outcomes (a misspelled word often goes undetected while a person reads a novel). One can easily sympathize with the neuroscientist who awaits a definition of neural information that could provide the kind of powerful predictive and explanatory link between local signal sequences and global system functions that is enjoyed by computer scientists as they work with artificial information-processing systems.

Cognitive neuroscientists may go beyond considerations of neural information processing to address the cognitive and behavioral consequences of the brain's computations, such as attending, believing, desiring, judging, deciding, finding food, or escaping a predator, all of which are generally useful to the individual's survival. While these cognitive and behavioral products of brain signaling can in principle be programmed into computer hardware of the appropriate design, other mental phenomena are believed to lie outside of the domain of the computer at the present time. Examples of mental activities that are believed to resist duplication in programming are understanding (e.g., Searle, 1980) and mental qualities, or "qualia," for example, perceived colors and aromas and feelings (Nagel, 1986; McGinn, 1991); others oppose this position (e.g., Churchland, 1986; Dennett, 1991).

The psychological approach to understanding attention, viewed here in terms of the question of "what is done" by attention to give the individual an adaptive advantage, leads one toward the topic of the goals of attention. In the following sections I describe both the adaptive goals served by attention and the manifestations of attention, or what it is that attention does that allows the individual to meet these goals.

## The Goals of Attention

Being able to pay attention has three major benefits for an individual: accuracy, speed, and maintenance of mental processing. These are defined in behavioral and cognitive terms, and in what follows

I frame the definitions in a general manner in order to encompass most of the attentional goals implied in the research literature.

*Accurate perceptual judgments and actions.* Accuracy in making a perceptual judgment or categorizing an object—correctly identifying an object, judging its aesthetic quality, and categorizing its color, orientation, size, distance away, or velocity of movement—is ordinarily not a problem for the individual when the object is the only item in the visual field. Difficulties sharply increase when other objects are in the vicinity, because information arising from distractors can confuse one's judgment of the target object, particularly when the distractors share features with the target. There is a large literature on the perceptual difficulties of searching for a target object in a visual display containing varying numbers of distractors of varying similarity to the target (e.g., Treisman and Gelade, 1980; Duncan and Humphreys, 1989). When objects are examined one by one in search tasks, it is assumed that attention restricts the range of incoming sensory information so that the information that is to be judged or identified arises from only the target object.

The accurate planning and performing of an action, whether that action be an external response or an internal mental operation, runs into difficulties when other, similar actions are available to the individual and when several separable actions or operations must be coordinated (e.g., Allport, 1989; Duncan, 1994; Pashler, 1992). When an individual is processing action information, attention may promote performance accuracy by preventing cross-talk between brain areas responsible for similar actions as an intended action is executed; this is similar to restricting sensory information when the goal of attention is increasing accuracy in perceptual judgments. Another goal, which currently is generating an increasing amount of research, is the organization of neural activity (e.g., Allport, 1989; Duncan, 1994) involved in the planning of actions.

When a mental process requires the organization or coordination of activities of more than one action component, attention is assumed to select an image of some kind (an image, e.g., of a goal of the action) to serve as an anchor around which the action components are organized or coordinated. An example of what might be called "organization anchoring" in planning to take an action is keeping in one's mind a sensorimotor image of how one would hold three large bags in two arms, or focusing on an intended meaning as one coordinated a sequence of words into an effective sentence. At the same time, attending to the goals of holding three bags or expressing an intended meaning protects the anchoring images from the interfering influences of related thoughts or actions (such as wondering which bag contains the milk you've just bought or writing key words concerning your talk on a blackboard). In view of these considerations, it would seem plausible to assume that attentional enhancement of an organizational anchor also contributes to the effectiveness of performance.

Thus, attention can increase the accuracy of perceptual judgments by selecting information flow on the input side of cognitive processing, and attention can increase the accuracy of actions on the output side of cognitive processing by selecting information flow in the organizing and planning of both internal and external actions.

*Speeded perceptual judgments and actions.* Attention increases the speed with which perceptual judgments and the planning/performance of actions takes place. When a stimulus is expected to occur, the expectation can direct attention to elevate activity at relevant brain sites prior to the actual occurrence of the stimulus, so that when the stimulus appears perceptual processing takes less time than it would had the stimulus not been expected. A driver can make a quicker start by anticipating a green light, and someone waiting for the phone to ring will be the first to answer it. In a somewhat similar manner, when an action is expected to be performed, this expectation can direct attention to elevate activity in brain sites that code the particular action, so that when the action is triggered it is performed more rapidly. For example, the accelerator pedal of an automobile will be depressed more quickly and with more force at the onset of a green light when the driver is attending to the feeling of tension in his or her foot. For more complex actions, the assembling of several action components can occur more quickly when an appropriate organizational anchor is brought to attention in advance. Examples are preparing to open a locked door with a key and preparing to reply to a lengthy question. As one prepares to use a key, an organizational anchor might be the composite image of the door moving inward as the right hand is turning the inserted key and the left hand turns and pushes the door knob. For someone about to compose a reply to a question, the organizational anchor could be the image or idea believed to be the main point of the question; the sooner the point is established, the faster the reply can be delivered.

*Sustained processing of a mental activity.* Attention allows us to sustain a perception or an action for extended periods of time even when the attention is not being driven by an expectation of a change in stimulus or a change in action. Typical examples of sustaining attention to sensory inputs are the continued enjoyment of pleasurable sensations produced by tasting food, listening to music, and watching a sunset, and examples of sustaining attention to action outputs are the repetitive humming of a melody, the contemplation of the amazingly diverse properties of the mathematical constant *e,* and the constant recycling of memories of events surrounding a recent tragedy.

The foregoing examples are intended to illustrate sustained attention to a process for its own sake. In this respect this third goal differs from the two other classes of attentional goals, accurate and speeded processing, because sustained processing apparently confers no immediate adaptive benefit to the individual. Possible remote benefits of sustained attention to pleasurable activities are elevated mood states that could promote more adaptive responses in someone faced with a challenging problem. For a great many people, however, the prospect of devoting prolonged attention to gratifying, esthetic, or contemplative experiences at the end of a workday motivates their toleration of routine and drudgery in their jobs.

Although many of the kinds of experiences that are frequently sustained for their own sake by attention are of the "qualia" type that are said to resist computational description, such as contemplating a happy memory for the feeling of contentment it brings (e.g., see Dennett, 1991, but see Johnson-Laird, 1988, for an opposing view), the goal of simply sustaining attention to any process can be computationally described in terms of the input-output relationships involved in the sustaining process.

Describing the goals of attention is the first important step toward understanding how attention operates, for these goals characterize the kinds of consequences or outputs that the attention process is presumed to produce for the individual. Although they are described at the rather coarse scale of behavior and cognition, goals can be evaluated at those levels in terms of their adaptive and experiential importance for a person, and awareness of the purpose of attention guides the steps researchers must take toward more precise descriptions of attentional processes at the level of brain pathways.

## The Manifestations of Attention

As we explore the attention process that produces the goals or outcomes described in the foregoing section, it will help us to have at hand a set of labels that characterize the ways that the attention process itself is manifested in behavior and cognitive experiences. Breathing, the movement of air in and out of the lungs, manifests itself not only by evident exchange of $O_2$ and $CO_2$ but also in the controlled exhalations that vibrate the vocal chords for speaking and singing. It is assumed in this book that attention manifests itself in three main ways that correspond rather closely to the three classes of goals just described. These three manifestations of attention— selection, preparation, and maintenance—will serve as organizing themes for the two chapters that follow, which deal with the data, concepts, and theories of attention obtained mainly from cognitive/ behavioral research.

*The selective manifestation of attention.* The goal of making a correct perceptual judgment of an object that is embedded in a cluster of objects requires that attention perform a selective operation on the flow of incoming sensory information. Specifically, the selective operation prevents information arising from distracting objects from entering the module that computes the particular judgment (that identifies the object, say, or detects its color). In like manner, the goal of correctly choosing one action from a set of alternative actions in working memory, as in the early stages of learning a skill, is related to the ability of attention to perform a selective operation on the information corresponding to the sets of actions. As perceptual and motor skills become well-learned, the manifestation of selective attention is presumed to decrease. But whether or not highly learned perceptual or motor skills ever become "completely automatic," in the sense of requiring no selective attention in processing them, is not clear (e.g., Shiffrin, 1988; Treisman and Gelade, 1980).

*The preparatory manifestation of attention.* The attentional goal of increasing the speed of perceptions and actions requires that attention be directed to a particular stimulus or action prior to the time that stimulus or action is expected to occur. When activity in the perceptual or action modules is already elevated at the time the stimulus or action occurs, the time to complete the perceptual or action processing is presumed to be reduced. William James (1890) described attentional preparation for a stimulus as "preprocessing" of the stimulus that reduced the amount of processing needed when the stimulus itself occurred.

Preparatory attention is assumed to be driven by an image of the stimulus or action that is held in working memory. Typically, working memory also holds a temporal expectation of when the stimulus or action will occur, so that the intensity of preparatory attention to the corresponding modules need not be increased to a high level until just before the onset of the stimulus or action. However, having expectations about the form and timing of perceptions and actions does not necessarily result in the buildup of preparatory attention in the relevant perceptual or action modules. The expectation in working memory may be held only in verbal form, as on the frequent occasions when we expect a green light to occur at a crossroad but do not form a strong image of it in anticipation.

*The maintenance manifestation of attention.* The sustained attention to a perception or action, with the continuation of an activity as its goal, is sometimes described as simply attending to something with "nothing else in mind." One important item that is absent from the mind in these cases is an expectation, held in working memory, of a related stimulus or action. While both preparatory and maintenance attention typically involve some degree of sustained attention, it is difficult to infer the presence of maintenance attention because most standard laboratory tasks entail clear behavioral goals that induce expectations in the subject. However, other measurements of mental processing are currently available to the cognitive neuroscientist, such as PET, functional magnetic resonance imaging (MRI), magnetoencephalograms (MEG), and electroencephalograms (EEG), and these brain-imaging techniques can provide objective indicators of maintenance attention when used in conjunction with suitably designed psychological "tasks" that can rule out the formation of expectations.

Although three different manifestations of attention have been described, it may be argued that all three involve the selection operation of attention. Preparatory attention involves selection of the particular stimulus or action that is prepared for, and maintenance attention is typically concentrated momentarily on a specific stimulus or action. The differences arise from the fact that some situations invoke selective attention that is sustained prior to an expected stimulus event and others invoke selective attention after the stimulus event has begun, and in some situations attention may be directed at an adaptive problem but in others it may not.

The three attentional manifestations may be more directly contrasted by describing the way they can occur in a trial of a typical psychological experiment. An example is the task in which the subject is asked to generate a verb that is associated with a noun (*apple* → *eat*). After a warning signal is given, the subject's attention is directed to and sustained at the location on the computer screen where the noun is expected to appear. Meanwhile, attention elevates the subject's readiness to think of the word and then say it. When the noun appears on the screen, it may evoke several associated words from which the subject selects an appropriate verb and utters it. After the response is made the subject typically turns attention to an evaluation of the correctness and possibly promptness of the response, especially if feedback is given, and may contemplate how he or she "feels" about the performance by sustaining attention in the maintenance mode.

## The Expression of Attention

The foregoing descriptions of the several goals and manifestations of attention led to the provisional claim that the selection process is present in all manifestations of attention. Earlier in this chapter another property of attention was described, that is, its ability to enhance activity in cortical circuits. This section considers the possibility that the selection and enhancement operations of attention arise from the same process.

*Attention expressed by algorithms operating on information flow.* It would seem that the appropriate way to give an information-processing description of how attention works—that is, how attention solves the problems of having too much information available to make a judgment or plan a response and of emphasizing a goal around which action components may be organized—is to state the algorithm or algorithms whose outputs do the job of selecting and enhancing an appropriate part of the information flow. Without referring to the type of hardware that embodies an algorithm of attention, one can infer the properties it would appear to possess. In general, the output of the algorithm would appear to modulate information flow rather than transform the actual content of the information. For example, computational models of attention (e.g., Cohen et al., 1990; Sandon, 1990) represent the selective attention process in this manner simply by changing the connection weights between particular sets of processing units within the main input-output pathway.

The specific form in which the modulatory output of an attention device is expressed in information-carrying elements can also be examined without referring to the instantiating device. The selection of a subset of information elements is produced by a difference between the modulatory processing of the target subset of elements and the processing of its surrounding elements. This difference can be produced three ways, and there exists a class of algorithms for each way: (1) by increasing the output at the target subset more than output in the surrounding subset is increased (the increase in this subset may be zero); (2) by decreasing the output at the surrounding subset less than it is decreased at the target subset (the decrease in this subset may be zero); (3) by both increasing the output at the target subset and decreasing the output at the surrounding subset (e.g., Keele and Neill, 1978; Neill and Westberry, 1987; Treisman and Sato, 1990). The first way of producing a selective difference in output at a target subset and its surround also elevates activity at the target site, thus generating both selection and enhancement properties of attention simultaneously. The combination of these two properties is particularly well illustrated by the preparatory manifestation of attention, in which the enhancement of the target subset is initiated prior to the onset of the stimulus and thereby speeds the processing of an object or action when it occurs.

Thus far we have discussed the modulatory output of classes of attentional algorithms without specifying what property of the information flow of neural signals is being modulated. The three properties of a neural discharge that can be modulated are its frequency, amplitude, and phase. Recordings of single neurons in the cortex of monkeys who are attending to the location or color of an object show changes in the frequency of neuron discharges (e.g., Moran and Desimone, 1985; Spitzer et al., 1988). Amplitude changes in neural discharges have not been systematically related to attention tasks but presumably could take place when neurons discharge in short "bursts," which can contain varying numbers of discharges in very rapid succession. Modulatory effects could also occur by changes in the phase or temporal onsets of the discharge signals, which can affect the synchronization of trains of discharges in separate pathways. The synchronization of discharges of neurons in separate cortical areas has been proposed as the means by which features of an object become bound into a unified percept (Eckhorn et al., 1988; Gray and Singer, 1989), and the conjoining of object features is assumed to be the main operation of attention, according to Treisman's well-known feature-integration theory (Treisman and Gelade, 1980). Of the three properties of a neural discharge considered here, it would appear that modulation by enhancement would be more directly expressed by increases in frequency or amplitude than by temporal changes in phase.

The foregoing analysis of possible modulatory properties of the attention algorithm is regarded as a computational analysis, because it concerns the output that is computed independent of the physical properties of the hardware that instantiates the computations. At this point one could proceed to program some version of the attention algorithm into a robot, and it can be shown that here one is faced with a choice among an almost limitless number of different algorithms that will generate the desired attentional outcome of selective processing. However, this book is concerned with the particular algorithm (or algorithms) by which the brain expresses attention. By examining what is known about neuroanatomy and neurophysiology, we not only substantially reduce the size of the set of potential algorithms that need to be examined, but we may also obtain some hints as to which algorithms to examine first.

Recent findings from neuroanatomy have advanced considerably our knowledge of how the many parts of the human brain are interconnected, and several different kinds of neurophysiological measures of brain activity in subjects engaged in attention-demanding tasks have identified specific brain areas that appear to be crucial for attentional processing. Given that the guiding purpose of attention research is the discovery of how attention operates to achieve its results, and given the prospect of a large set of potential algorithms that could deliver the desired attentional outputs, it appears that we must first find the brain mechanisms in which the attention algorithms are instantiated if we are ever to determine the exact form of the algorithms.

## Plan of the Book

My presentation of current research on attention has two main parts. The first part (Chapters 2–3) addresses the question of what attention does for the organism, and this part of the story is based on cognitive analyses and behavioral responses to psychological tasks that are specifically designed to manipulate attentional processing. The second part (Chapters 4–5) examines the neuroanatomy of the brain pathways presumed to be involved in attentional processing and reviews some of the major physiological studies of neural activity in these pathways in subjects performing attention tasks. The hypothesis is examined that two subcortical structures, the superior colliculus and the thalamus, may contain circuits that embody algorithms of attention. The anatomical connections between these two subcortical areas and other relevant brain areas are reviewed, and since the connectivity of the thalamic circuitry is relatively well established, it is described in some detail. This is followed by computer simulations of a set of neural networks based on several versions of the thalamic circuit, in which it is shown that all of these network versions produce an output pattern that fits the output pattern of one class of attention algorithms, the class that produces a difference in information flow at the attended site and that at the non-attended sites chiefly by enhancing the flow at the attended site.

After these two descriptive parts, I present in Chapter 6 a cognitive-neuroscience model of visual attention in an attempt to synthesize the major empirical and analytic results that the book surveys. Finally, I propose four general principles of attention as a basis for building cognitive-neuroscience theories of attention in Chapter 7.

While the overall goal of the book is to understand the workings of attention in the domains of action as well as in the domains of perception, the vast majority of cognitive and neuroscience research has been concerned with attention to visual objects and their attributes. Hence, this book will draw heavily upon experiments and theories of attention as it has been viewed as a modulator of visual sensory inputs. It is hoped that this knowledge can provide appropriate material from which cognitive-neuroscience principles of attention can be formulated and then generalized to the expressions of attention to other sensory modalities and to the cognitive domains where motor plans and ideas are shaped by intellectual actions.

The question of how attention accomplishes its effects in brain pathways is clearly not yet fully answered. The approach taken here, by combining research methods and concepts from the several disciplines of cognitive psychology, neuroscience, and computer science, has provided substantial progress toward that answer, however, and I hope it will continue to do so in the future.

# 2

## Selective Attention

The manifestations of attention named in Chapter 1—selection, preparation, and maintenance—encompass many of the ways that attention has been described in the cognitive-psychological literature. The literature treats not only objective experimental indicators of attention but also subjective experiential indicators of attention, and both indicators have contributed to our present cognitive views of the field. Although the three manifestations may be combined in many situations of daily life and of laboratory experiments, most experimental and theoretical accounts, including this one, emphasize only one at a time. It may be helpful to readers familiar with contemporary attention research to indicate here that the processes of orienting (e.g., Posner and Petersen, 1990) will be treated in the next chapter, on preparatory attention, and the processes of search (e.g., Shiffrin and Schneider, 1977; Treisman and Sato, 1990) will be addressed in the present chapter.

### Early Theories of Selective Attention

The earliest influential treatment of selective attention came from the pen of William James, in his chapter on attention in *Principles of Psychology*. James emphasized selectivity as the hallmark of the attention process: attention takes "possession of the mind, in clear and vivid form, of one out of what seems several simultaneous possible objects or trains of thought. Focalization, concentration of consciousness are of its essence. It implies withdrawal from some things in order to deal effectively with others" (1890, vol. 1, p. 403).

Toward the end of the nineteenth century, psychology was just beginning to gain an identity as an academic discipline separate from philosophy, and it could be said that James's penetrating treatment of the attention process helped distinguish the two fields. Philosophy had apparently avoided the term *attention* (Kant briefly treated attention only in a footnote in the *Critique of Pure Reason* of 1787), because the process had almost always been considered an aspect of consciousness. James was the first well-known philosopher to treat it separately. In the opening pages of his chapter on attention, James pointed out that the traditional empiricist position of philosophy regarded the person as "absolutely passive clay, upon which 'experience' rains downs" (p. 403). In contrast, he claimed, "each of us literally *chooses,* by his ways of attending to things, what sort of universe he shall appear to himself to inhabit" (p. 424). For James, attention is of central importance in determining the particular experiences of a person: "My experience is what I agree to attend to" (p. 402).

The stream of experience provided James with most of the raw material for his academic contributions. James, like his brother Henry, the renowned novelist, had a gift for describing the stream of experience in a sensitive and clear manner that invited the reader to observe the richness of his or her own experiences in a more vivid manner. A glance at almost any page of his written works attests to the power of his self-observations. It was his own stream of experience, highlighted at any moment by the focus of his attention, that furnished James with the bulk of the ingredients for his theories, both psychological and philosophical.

When James offered specific "physiological" definitions of attention (p. 434), he characterized it as "the accommodation or adjustment of the sensory organs" or the "anticipatory preparation from within of the ideational centers concerned with the object to which the attention is paid." On the surface, the first definition seems more restrictive than the somewhat broader notion of selection described in this book, while the second definition seems to correspond more closely to the preparatory manifestation of attention. The "accommodation or adjustment of sensory organs" implies that "any object, if *immediately* exciting, causes a reflex accommodation of the sense-organ, and this has two results, first, the object's increase in clearness; and second the feeling of activity in question" (p. 435). The phrase "increase in clearness" is a distinctly cognitive type of description (as contrasted with a neurobiological description), but it captures the computational flavor of the selective aspect, since clearness implies a decrease in the confusing inputs from other objects, that is, less noise or ambiguity in the information stream (for making judgments of identification, for example, or for detecting color). As a result of a decrease in irrelevant input, the outcome of the processing will be a more adaptive response to the sensory input.

But James applied the processes of "accommodation or adjustment" to the internal domain of ideas as well as to sensory objects, by maintaining that such a mechanism assists in the concentration of attention to the idea of an object. The act of trying to remember something was said to be accompanied by a "feeling of actually rolling outwards and upwards of the eyeballs, such as occurs in sleep, and is the exact opposite of their behavior when we look at a physical thing" (p. 436). James spoke of selecting ideas on the basis of their interest to the person, ideas which then took possession of the mind, much in the same way as he spoke of accommodating the senses to objects.

By 1920, ten years after James died, behaviorism imposed a blackout on much academic discussion of unobservable entities such as ideas and attention, particularly among psychologists. For thirty years the ideal explanation in psychology was to account for a response solely as a function of the stimulus input. In the early 1950s, however, both empirical and theoretical studies began to suggest that the stimulus input was subject to some kind of selective manipulation, and that this manipulation came from processes within the organism. Lawrence (1950) published a pair of studies on the acquired distinctiveness of cues, in which it was shown that rats in a T maze can be trained to respond to one part of a stimulus under one condition and to ignore that part and respond to another part of the stimulus under another condition. The implication of these results was that a selected part of sensory input controlled the response, and that this selection process could be modified by learning.

At about the same time that the phenomenon of perceptual selectivity was being substantiated by experiments, Estes (1950, 1954) proposed a theoretical way to describe selective operations during stimulus reception. His "stimulus-sampling" notion is based on the simple assumption that the organism processes only a part of the available stimulus elements but that the subset that is sampled from occasion to occasion (or from trial to trial) fluctuates in a random fashion. The stimulus-sampling assumption gives rise to a mathematical expression of the response probability on a particular trial and a mathematical expression of the change of learning that occurred on that trial. The learning process is represented in stimulus-sampling theory as the change in connections of the sampled elements to the available responses. The sampling assumption provides a mathematically elegant and manageable way of showing how learning can proceed by small increments on each trial. By assuming that the sampling of stimulus elements is random, the mathematical equations expressing both response probability and learning can be written in very simple forms. The linear difference equation describing trial-by-trial learning became a standard fixture in a host of stimulus-sampling models published over the next decade, and it is still being used (e.g., Healy et al., 1992; LaBerge, 1992, 1994).

Perhaps the major difference between the stimulus-sampling idea of Estes and the traditional stimulus-selection idea in attention lies in the matter of what controls the selection process. In Estes's theory, the selection of stimulus elements is random, and this simple assumption enabled mathematical models to predict the detailed data from a wide variety of learning tasks. In theories of attention, it is presumed that selection of stimulus information is not random but rather determined by properties of the stimulus and by internal states of the organism. There have been efforts, however, to modify the random property of the stimulus-sampling assumption to fit attentional tasks (LaBerge, 1994; Prokasy, 1961).

Soon after Estes's work was published, Guthrie (1959) modified his previous behaviorist learning theory to allow for a selection process that strongly resembled selective attention. In his words, "What is noticed becomes the stimulus for what is done." Thus, by the end of the 1950s the behaviorist approach that had effectively shut off serious consideration of attention for thirty years was giving the selective manifestation a central role in the stimulus-response paradigm, the backbone of behaviorist learning theory.

But it was from outside the strong learning-theory movement of the 1930s, 1940s, and 1950s that selective attention received its most formidable advance in the field of psychology. In the highly influential book *Perception and Communication* (Broadbent, 1958), Broadbent proposed a theory of attention based on notions of information theory in the communication of messages. When a particular spoken message is delivered in the company of other spoken messages, such as can occur in a conference telephone call or at a social gathering, the listener receives the target message if the total array of incoming information can be filtered and only the appropriate information allowed to pass. The listener must, of course, direct the filter, or "tune" it, to the "channel" over which the target message is sent. Earlier experiments by Cherry (1953) and Broadbent and Gregory (1964) investigated selective listening in "shadowing" experiments, in which separate spoken messages are presented simultaneously by earphones to the two ears and attention was engaged to one message by having the subject repeat aloud that message. The expected result, on the assumption that attending to one message would eliminate perception of the other, is that no information from the other channel would be responded to or remembered. Stated in terms of Estes' sampling notion, information elements from the non-shadowed (non-attended) channel would not be contained in the sampled set and therefore would affect neither momentary response probability nor learning.

The strong assumption of the eliminative filter began to be called into question by experiments that showed that under certain conditions some of the information on the rejected channel is sampled and processed. Moray (1959) showed that subjects noticed that their own name was uttered on the rejected channel, and they could remember some instructions delivered on that channel following the occurrence of their name. Treisman (1960, 1964a,b) found that subjects would often switch attention from one ear to the other to follow a sentence when the messages to the two ears were abruptly interchanged. Apparently individuals process highly meaningful words outside the attended "channel," and this finding led Moray (1959) and Deutsch and Deutsch (1963) to make the strong assumption that all information in the input stream is perceptually processed and that selection occurs at a stage following perception. A problem with this position is that it does not easily account for the fact that messages on rejected channels are so strongly reduced.

An alternative to the "late selection" account is to maintain the notion of "early selection" but assume that the filter is not completely eliminative. Treisman (1960) proposed that information flowing in unattended channels is not switched off but simply weakened or attenuated. Words of greater significance to the subject have lower perceptual thresholds and therefore perception of these words is more likely to be activated when their input information is at a low signal-to-noise ratio. It should be noted that signal-to-noise ratio of information is not necessarily the same as the physical intensity of the stimulus, such as volume. The fact that we can attend to the voice of a quiet talker standing next to a loud talker suggests that the information from the loud talker that is attenuated is located upstream from the primary auditory area of the cerebral cortex but still within the "early" perceptual stages of auditory processing.

Today, most researchers agree that late selection can take place on the basis of the individual's momentary goals and response tendencies. What is still controversial is whether or not early selection occurs at all. One of the clearest ways to answer this question is to observe activity in the brain while an individual is engaged in attention tasks, so as to determine both the temporal and spatial markers of this activity. Detailed treatment of these neurobiological experiments will be postponed to the second half of this book, but it can be said here that scalp-recorded event-related electrical potentials (ERPs) and event-related magnetic fields (ERFs) show attention effects within the first 100 msec following a visual or auditory stimulus and these effects are located in brain areas near the sensory projection area of the stimulus (e.g., Heinze et al., 1994; Woldorff et al., 1993). The location of attention effects in early sensory areas of the brain have been generally confirmed by PET studies (e.g., Corbetta et al., 1991).

The notion of a filter as the mechanism that selects information has the distinct implication that the selection occurs by the inhibition or blocking of information surrounding the information arising from the target source. For example, a kitchen sieve or an acoustical filter in the physics laboratory merely passes along the selected material or signal without changing its amplitude or gain while blocking other material or signals. Thus, when the filter metaphor is exported to the field of attention, it is assumed that the expression of attention is an output of an algorithm that favors information coming from a target over the surrounding information solely by inhibiting information in the surround.

## Selection of What, Where, and Which

In most research, the attended "site" within an internal brain map corresponds to a spatial region of the visual field. For example, the spotlight metaphor describes (depicts) the attended area as a single connected region of an internal map that exhibits higher activity than the activity in the surrounding area. Internal maps are presumed to represent attributes of objects as well, such as color, line orientations, and movement directions. Even though locations and attributes may be coded coarsely in large receptive fields and/or in a distributed manner in many areas in some brain maps (Chapter 4 will discuss this issue in more detail), the maps are presumed to encode similarities between attributes as well as distances between locations. Therefore, considerations here of the properties of the attended site within a location map are intended to generalize to attended sites within an attribute map.

Recently, several researchers have suggested that selective attention also operates on objects (Baylis and Driver, 1992: Duncan, 1984; Egly et al., 1994; Kramer and Jacobson, 1991; Vecera and Farah, 1992) or object tokens (Kanwisher and Driver, 1992). The concept of an object token, proposed by Kahneman and Treisman (1984) and developed further in a later paper (Kahneman, Treisman, and Gibbs, 1992), represents the combination of properties exhibited by a displayed object that are stored in episodic memory somewhat like a file folder contains documents describing the various aspects of a particular topic. When two objects share the same attribute, as a rose and an apple share the color attribute red, each shape is combined with its color and location codes in an object file or object token. Therefore, one can speak in terms of "which" object as well as "what" object or "where" the object is located.

When two objects are presented simultaneously and attention is directed to one of them, it is conceivable that the expression of attention to the object token could take place in a specific cortical brain area that specializes in processing object tokens, in the same way that expression of attention is assumed to take place in posterior (cortical) parietal areas for location and in inferior (cortical) temporal areas for shape attributes. Kanwisher and Driver (1992) suggest that the combining of the what, where, and which of an object token may take place in the anterior superior temporal area. If an object token operates as a memory representation in a prefrontal area or areas, then activity in these anterior cortical areas could act as a control on attentional expression in the location and attribute processes in posterior cortical areas.

Single-cell recordings in monkeys (Goldman-Rakic, Chafee, and Friedman, 1993) have shown that neurons in the prefrontal cortex and parietal cortex respond similarly during time delays in which spatial locations must be remembered. A PET study by Jonides et al. (1993) has shown increased blood flow in these two areas during a spatial working-memory task in humans. Working memory for object shape and color has been shown to activate neurons in the monkey prefrontal area just below (ventral to) the spatial working-memory area. Wilson et al. (1993) recorded from a prefrontal area in which neurons responded preferentially to delays in cued locations and cued attributes such as color and shape. Additional research with PET scans and single-cell recordings would help us decide whether the processing of object tokens occurs in both anterior and posterior cortical areas, or whether it is largely confined to anterior areas.

## Six Properties of the Attended Area

The inferred properties of the selected area of attention have inspired a rich and varied collection of metaphors, most of which I mentioned in Chapter 1: attention has been seen as a filter, effort, resources, a control process of short-term memory, orienting, conjoining object attributes, a moving spotlight, a gate, a zoom lens, a channel, and a distribution of activity. Examination of the properties of the selected area of attention is aided by studies that use tasks involving the identification of visual shapes and their attributes, because visual shapes can delineate the metrics of visual space relatively precisely, particularly when the shapes are small. Furthermore, the identification process is assumed to require more careful attention to the characteristics of a shape or object than simply detecting its presence or absence. It goes without saying, however, that identification tasks will be useful for present purposes only if the perceptual processing involves attention.

Many studies, particularly in the search literature, have explored the characteristics of stimulus displays and instructions that induce attentional processing as opposed to pre-attentive "pop-out" of a shape or attribute (e.g., Treisman and Gelade, 1980; Julesz and Bergen, 1983). Also, in the domain of ideational processing, it is presumed that tasks differ in the amount of attention required; for example, for a subject generating a verb to match a presented noun, the level of attention given to the process decreases as the same list of nouns is repeated (e.g., Raichle, 1994). In the following sections, references to "the attended area" during perceptual or ideational processing presume that task conditions have been met that induce attentional processing.

*The boundaries of the attended area.* A boundary typically implies that there exists a relatively sharp transition from one area or domain to another. Thus, in the case of a bounded area of attention, the transition from the attended area to its surround is presumed to be relatively abrupt, analogous to the sharply defined beam of a well-focused spotlight or searchlight. When attention is directed to an object or objects in a field containing many objects, it is plausible to assume that the boundary of the attended area falls between objects and seldom cuts across an object. However, a variant of the spotlight metaphor, the zoom-lens model (Eriksen and St. James, 1986), depicts the attended area as having a high clarity at the center and a gradual decrease in clarity with distance from the center. Thus the periphery of the attended area resembles the tapering off of a fringe. Under a very high power setting, however, the clarity of attentional processing could drop off so rapidly that the effect is one of an abrupt boundary. But at the typical settings, the zoom lens creates a gradual reduction of attentional processing with distance from the center, rendering the notion of a boundary less useful.

An experiment from our laboratory was designed to reveal the existence of a relatively sharp boundary between an attended object and surrounding objects (LaBerge et al., 1991). Human subjects were instructed to identify a letter at the center of a chain of seventeen letters (e.g., the letter C embedded in H's). The center (target) letter was randomly selected from a set of four letters, two of which (C, H) were assigned to a righthand button and the other two (S, K) to a lefthand button. Results obtained in a similar design by Eriksen and Eriksen (1974) led to the expectation that the response time to the target C would be greater when its distracting letters were S or K than when its distracting letters were H because a KCK type of display contained conflicting information concerning the correct response (the target, requiring a righthand response, is surrounded by stimuli requiring a lefthand response) while an HCH type of display contained only redundant information (both target and surround would require a righthand response). However, if the attention boundary could be positioned between the center letter and its closest distracting neighbor, then the time to identify and respond to the center letter would not be affected by the type of distractor, since the information from the distractor lay outside the area of attention.

The measure that indicates whether or not information from the surrounding shapes is getting through to the response is the difference in mean response time *(RT)* between the display containing incompatible *(I)* response information and the display containing compatible *(C)* response information. For convenience this measure is called $I - C$. Eriksen and Eriksen (1974) had obtained an $I - C$ value of approximately 80 msec, when the separation between the target and its adjacent item was 0.06 degrees, indicating that the attentional area of the subjects included adjacent distractors, despite explicit instructions to ignore them. Our goal was to find a way to concentrate attention exclusively to the target item, that is, to show that a sharp attentional boundary could be placed between a target and its closest flanking objects.

To attempt to reduce the size of the attended area around a letter item, we used the same method that previously had been used to change the width of attention from one letter to a five-letter word (LaBerge, 1983). We presented another string of items just prior to the string containing the C, H, S, and K items. The prior display required the subject to identify the digit 7 at the center of a string of letters with similar shapes (T, Z), thereby concentrating the subject's attention at the same location as the C, H, S, or K target item in the second string. When the exposure time of the display 7 was varied from 450 msec to 67 msec, the value of $I - C$ changed from approximately 22 msec to 9 msec. The largest $I - C$ value, 22 msec, is considerably less than the 80 msec value obtained by Eriksen and Eriksen (1974), but our inter-item separation was 0.14 degrees (versus their 0.06 degrees), and their target display was presented for a much longer period (1,000 msec versus our 200 msec) and was un-masked, while ours was followed by a mask display. To attempt to reduce the $I - C$ value to zero, we asked four dedicated subjects to run 24 sessions over 12 days; 12 sessions were run at the shortest exposure time in which they could identify the digit 7 on at least 85 percent of the trials, and 12 sessions were run at a longer exposure time that added 200 msec to their minimum value (the minimums were 50 msec for three subjects and 33 msec for one subject). The resulting $I - C$ values at the minimum duration of the digit-7 string were 0, 0, 1, and 2 msec, while the $I - C$ values at the longer dura-tions were 15, −2, 13, and 5 msec, respectively. Thus, for the four highly trained subjects, the influences of the distracting items on the responses to the target item were virtually eliminated. This find-ing would seem to support the conclusion that the attentional boundary for these subjects was located between the target item and its adjacent item, which means that the diameter of the attended area was less than the half-degree of the target-letter region between the two distractors.

When distractors are similar to the targets, as were the letters C, H, S, and K in these experiments, it is assumed that the attended area must be restricted to the target object long enough for a correct identification to take place. But experiments have shown that it is fairly difficult to obtain $I - C$ evidence to support this. Much more typical is the finding that $I - C$ is well above zero, indicating that the attended area has somehow included at least one distracting item in addition to the target item. But, if one assumes that attention allows only one shape to be identified at a time in this situation, where distractors are similar and close to the target, what is happen-ing to the attended area during the short time that the display of letters is on?

A plausible answer is that the attended area fluctuates rapidly while its size is made to conform to that of a single object. Sometimes the attended area shifts to an object flanking the target, but most often it is aligned on the target item. When an string of items is very briefly exposed, as in the case of the digit-7 string in the experiment just described, the fluctuation must be constrained, and this con-straint keeps the attended area aligned at the location of the center item when the subsequent letter string appears. Thus, an effective attention width of several items, upon careful analysis, may turn out

to be a width of rapid fluctuation of a smaller attention width corresponding to one item. A mathematical model of attention fluctuation in this type of experimental task has been described elsewhere
(LaBerge, 1992, 1994).

But if one relaxes the "one-at-a-time" assumption in the identification of objects or attributes, then the influence of neighboring
distractors on the identification of a target can be accounted for by
assuming an enlargement of the attended area so that it encompasses more than one letter. The zoom-lens metaphor of attention
allows the width of the clearly perceived area to be varied, and when
this area includes several shapes, these shapes can be processed simultaneously (Eriksen and Yeh, 1985). The optimal setting of such
a zoom lens is as wide as would be needed to identify as many objects
and attributes as possible. But at wide settings the resolution of detail
is less than it is at narrow settings. When identification of a shape
requires fine discriminations, therefore, the zoom lens must be narrowed, and in some display conditions the narrowed lens may resolve
only one object (Ericksen and Webb, 1989).

A serious problem for testing this theory is to determine independently the size of the lens in any given display condition. In the case
of the $I - C$ experiment just described, the similarities and proximi-
ties of the targets and distractors may require the zoom lens to con-
tract to a size that encompasses only one letter at a time. To account
for $I - C$ values greater than zero (values that indicate processing
of the distracting object), the zoom lens would be required to move
from one letter location to another in the same way as the zoom
lens has been assumed to scan items one at a time in a search task
(Eriksen and Webb, 1989). Thus at a sufficiently narrow setting, the
zoom lens is assumed to create an attended area having a sufficiently
sharp boundary to allow processing of a target letter in the $I - C$
task while blocking the processing of a neighboring distractor.

The zoom-lens metaphor implies that the movement of the attended area between object positions takes place in an analog manner, much as a physical lens would be moved from one location to
the other. The movement of attended area will be considered later
in this section.

*The variable size of the attended area.* Jonides (1983) described a
model of visual attention in which attention could be set either
widely across a display of objects or narrowly to the size of a single

object. Both the dispersed and focused settings of attention assumed a uniform distribution of attention within the attended area. The wide setting of attention is appropriate for parallel processing of items, as in search displays containing a single item that "pops out" pre-attentively. The narrow setting of attention was presumed to be preferable for serial processing of items, as in search displays in which the objects are constructed by conjunctions of attributes (e.g., searching for a blue triangle in a display of blue squares and yellow triangles). Eriksen's zoom-lens model (Eriksen and Yeh, 1985) can be viewed as a modification of Jonides's two-process model, in which the distribution of attention may be adjusted continuously between a wide and narrow setting. However, the zoom-lens modification carries with it the implication that attention operates in a manner that clarifies or resolves details within the attended area, while the two-process model simply assumes that attentional "resources" are contained within the attended area.

In another study (LaBerge, 1983), the size of the attended area was measured by presenting a probe stimulus in one of five locations along a horizontal line in the visual field. The time required to identify that stimulus was taken as the indicator of how strong attentional processing was at that location. This experiment was replicated with six trained subjects (LaBerge and Brown, 1989). The size of attention was varied from the width of one letter (about 0.30 degrees of visual angle) to the width of five letters (about 2 degrees of visual angle) by presenting a single letter or a five-letter word to be quickly identified just prior to the presentation of the probe (see Tables 2.1A and 2.1B). The response-time data are shown in Figure 2.1.

The V-shaped curve in Figure 2.1 indicates that attention was concentrated at the location of the center target item (a single letter) just prior to the probe stimulus, while the approximately horizontal curve indicates that attention was extended rather uniformly across the larger target, the five letters of a word. Subjects displaying wide attention (to a word) showed lower overall response times than subjects displaying narrow attention (to a letter), which suggests that in this experiment the amount of attention dedicated to the location of a letter is greater when one is attending to the whole word than only to that letter. In this study the width of the word was approximately 2 degrees. When attention is spread more widely, over 3.8 degrees (Egeth, 1977) or 3.6 degrees (LaBerge and Brown, 1989),

*Table 2.1A*

| | |
|---|---|
| Warning signal | ✳ ✳ ✳ # ✳ ✳ ✳ |
| Target #1 | 8 5 8 S 5 8 5 |
| Target #2 | I R I |
| Locations of target #2 | ● ● ● ● ● |

Catch trials for S were 5, 8
Catch trials for R were P, Q

Instructions: Press the button only when S
  is followed by R

*Table 2.1B*

| | |
|---|---|
| Warning signal | ✳ # # # # # ✳ |
| Target #1 | ✳ B R U C E ✳ |
| Target #2 | I R I |
| Locations of target #2 | ● ● ● ● ● |

Catch trials for BRUCE, RALPH, etc.,
  were TABLE, CHAIR, etc.

Catch trials for R were P, Q

Instructions: Press the button only
  when a given name is followed by R

however, the overall response time to a probe in a letter location is greater than when attention is spread narrowly. These findings are not surprising, since one might expect that the amount of attentional concentration at a given point would approach a limit as the total concentration is spread more widely.

When attention shifts from a whole object to a part of the object, the size and often the location of the attended area is changed, and the rate of change appears to be quite rapid. When an object is presented to our eyes, the abrupt and salient luminance change at its borders induces the attended area first to encompass the whole object, providing that the object is not too large. But almost immediately the size of the attended area is scaled down as parts of the object are rapidly examined (see Bundesen et al., 1984; Olshausen

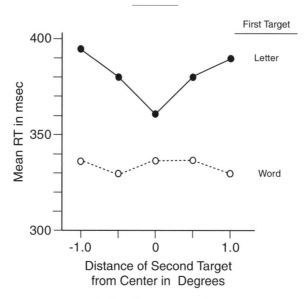

*Figure 2.1.* Indicators of narrow *(upper curve)* and wide *(lower curve)* areas of preparatory spatial attention. The data points represent mean response time (RT) involved in identifying a second target after the subject had already identified a first target (either a single letter at the center of the field). The second target |R|, was located either in the center or in one of two locations to the left or right of center. The trial displays are shown in Tables 2.1A and 2.1B.

et al., 1993). William James once commented (as quoted in Myers, 1986, p. 186) that the whole-then-part sequence of attending was largely involuntary. Whether such involuntary perceptual analysis is natively determined or learned after an exceedingly large number of experiences with a variety of objects is not yet known.

*The variable intensity of the attended area.* For a spotlight of a given size, the beam of light can vary from low intensity to high. Does this property carry over metaphorically to the notion of an area of attention? The zoom-lens model is based on the idea of changing the focus of the lens, which varies the resolution of an attended area but does not change the overall intensity. Most simple spotlight metaphors of attention have tacitly assumed a constant intensity within a moveable attended area (e.g., Tsal, 1983; Posner, 1980). Psychologists who regard attention as the allocation of resources are more likely to suppose an attended area of a particular size with varying amounts of resources, interpretable as varying intensities of attention. A more extended discussion of attentional resource models will be given in the next chapter, which is concerned with the preparatory manifestation of attention.

*Attending to one connected area at a time.* The notion that we attend to only one thing at a time is deeply rooted in our culture. William James (1890, vol. 1, p. 409) expressed this belief in his characteristically clear and vivid style:

> If, then, by the original question, how many ideas or things can we attend to at once, be meant how many entirely disconnected systems or processes of conception can go on simultaneously, the answer is, not easily more than one, unless the processes are very habitual; but then two or even three, without very much oscillation of the attention. Where, however, the processes are less automatic, as in the story of Julius Caesar dictating four letters whilst he writes a fifth, there must be a rapid oscillation of the mind from one to the next, and no consequent gain of time.

In the years since James wrote these words, the one-at-a-time property of attention has been regarded in information-processing terms as a "bottleneck" (e.g., Welford, 1952; Broadbent, 1958) and has been investigated by Pashler (1992, 1994) in the context of the interference produced by performing two tasks at the same time.

The "one-at-a-time" processing of attention is an assumption that arises more "naturally" from a spotlight metaphor than perhaps from any other metaphor of attention. After all, a spotlight can illuminate only one area at a time, though it can take a variety of shapes. Perhaps more significant for the purposes of our metaphor is the fact that when a spotlight moves away from a location, that location is immediately plunged into darkness.

Other metaphors that have embodied the "one-at-a-time" processing assumption are gates or channels. These, on the other hand, imply a different mechanism for shifting attention across the visual field: when attention shifts from one location to another it simply opens a new gate or channel while the old one closes. Of course, the gate and spotlight metaphors become more similar if one considers that a beam of light turned off and on will seem as discrete as the opening of two different gates or channels. However, the time needed to shift attention will increase with the distance moved in the case of the spotlight but not necessarily in the case of the channel or gate.

Another difference between the spotlight and gate metaphors is that a gate-like attention area would not be strictly limited to the "one-at-a-time" processing assumption. When a new gate or channel is opened at another location, the closing of the original gate or channel need not occur at the same time, thus allowing for the possibility that, however briefly, two gates are open. If attention is shifted sufficiently rapidly from one location to another, it is possible that several gates may be open simultaneously. A hypothetical example is the simultaneous activity of several items in working memory induced when one briefly attends to each item in rapid succession.

*The movement or shift of the attended area.* Of all the properties of selective attention, that of movement has been the most controversial in the recent literature. Many who subscribe to the spotlight metaphor of attention for purposes of conceptualizing a bounded area of attention extend the metaphor to the way a physical spotlight moves, that is, in a continuous or analog manner across space. Those who consider orienting as a prototypical manifestation of attention (e.g., Posner, 1980) have usually been led by notions of how the eye moves in space. Considered the "eyeball behind the eyeball," attentional orienting is viewed as a covert analog of overt eye movements, and therefore movements of attention can be regarded as sharing characteristics with movements of the eyes. Furthermore, the brain areas that are involved in eye movements (e.g., the superior colliculus and posterior parietal cortex) have been implicated in the processes of covert orienting (e.g., Johnson, 1994; Klein, 1980; Posner, 1984; Rizzolatti et al., 1987). The discussion and review of evidence relating to orienting and analog shifts of attention will be postponed to the next chapter, where they will be treated under the general category of preparatory attention.

*The duration of attention.* When a selected area or site in brain pathways expresses attention, how long does it typically remain active? Apparently only very briefly during visual search or skilled reading, in which less than 100 msec are needed to selectively sample information from an object in a visual display (e.g., Treisman and Gelade, 1980); and in the case of serially scanning items of memory, on the order of 25 msec are required (Sternberg, 1966). The rapid scanning of parts of an object would seem to involve momentary snap-shots of selective attention having similarly brief durations. During periods of preparatory activity prior to the occurrence of an expected target, however, the expression of attention at the target site may be prolonged well beyond the fraction of a second that simple selective attention can operate. Sustaining attention for relatively extended periods of time would seem particularly important for obtaining measurements of attention in brain-imaging experiments. Estimates of the duration of preparatory attention depend upon whether preparation is defined strictly as continuous, uninterrupted preparatory activity at a target site or more loosely, as in the case in which the expectation of an impending target repeatedly directs preparatory attention to a target site that briefly wanders (the important distinction between an expectation and an attentional preparation will be discussed in the next chapter). Furthermore, when considering the duration of attention one usually wishes to consider what level or intensity of attentional activity is involved.

Experiential observations of the duration of continuous attentional activity have never been described better than by William James (1890, vol. 1, p. 420). He was convinced that attention is a fleeting activity: "There is no such thing as voluntary attention sustained for more than a few seconds at a time"; "what is called sustained voluntary attention is a repetition of successive efforts which bring back the topic to the mind." What seems to help the sustaining of attention is in the stimulus itself. "No one can possibly attend continuously to an object that does not change." And if the stimulus or idea is unchanging, then the "*condito sine qua non* of sustained attention to a given topic of thought is that we should roll it over and over incessantly and consider different aspects and relations of it in turn."

Helmholtz (quoted in James, 1980, vol. 1, p. 422) spent more time in the laboratory than did James, and his experiences there led him to the same general conclusions.

An equilibrium of the attention, persistent for any length of time, is under no circumstances attainable. The natural tendency of attention when left to itself is to wander to ever new things; and so soon as the interest of its object is over, so soon as nothing new is to be noticed there, it passes, in spite of our will, to something else. If we wish to keep it upon one and the same object, we must seek constantly to find out something new about the latter, especially if other powerful impressions are attracting us away.

Thus both James and Helmholtz viewed attention as a fluctuating, volatile activity that is briefly held in place by a combination of changes in the stimulus itself and voluntary control that is sustained by a person's interest in the stimulus.

Objective experimental observations of the duration of sustained attention have generally supported the view that attention can be sustained only for a matter of seconds, but the data are not as precise nor as consistent as one would like, owing to the complexities involved in designing even the simplest experimental tasks. Apparently the most sensitive measure of the duration of preparatory attention is response time to the appearance or change in a stimulus, when the stimulus is presented randomly within some range of time intervals following a warning signal. The time between the onset of a warning signal and the onset of a target stimulus is termed the foreperiod, and the duration of attention is measured by the range of foreperiods over which response time remains at its shortest values.

One of the pitfalls of measuring attentional activity over a range of foreperiods is that the expectancy of a target may not be constant over the range. For example, when a stimulus may occur with equal probability at times between 1 and 4 seconds, the expectation of the target increases as the foreperiod approaches 4 seconds. As the expectation increases, the corresponding attentional preparation is presumed to increase, not remain constant over the foreperiod range. Experiments that allow expectation to grow in this manner have consistently shown an increase in response time with foreperiod duration (for a review, see Niemi and Naatanen, 1981).

The most effective experimental procedure for maintaining a constant expectancy for a target over a time interval is to distribute the target onsets not as a rectangular distribution (equal foreperiod probabilities) but as an exponential distribution (decreasing foreperiod probabilities), in which the probability of a stimulus occurring given that it has not appeared up to that point is equal throughout the range of foreperiods. (In other words, the hazard function of stimulus occurrence is constant.) For a clear and thorough discussion of the foreperiod issue and response times, the reader is referred to Luce (1986). Data gathered in experiments delivering stimuli of moderate intensity with exponential foreperiod distributions show flat mean response times between 500 and 1,500 msec in one study (Green and Luce, 1971), and between 200 and 800 msec in another study (Green et al., 1983). In the latter, the foreperiod range was 0 to 6.5 sec, suggesting that attention can be sustained at a high and constant level only for durations lasting about a second. Furthermore, the flat response-time portion of the foreperiod range appears to depend upon the level of attention required to perform a task. Green and Luce (1971) found that lowering the intensity of a stimulus shortened the flat region of the response-time curve while shifting the curve upward.

Although further systematic investigation of the time limits of sustained attention are called for, not only in the auditory modality but also in the visual modality, it appears that the duration of continuous attention, whether of the preparatory type or the simple selective type, is apparently confined to a very short range, as observed experientially by James and Helmholtz. The computational question arises as to why human and animal systems are designed with such limits on the sustaining of attention. One might conjecture that there are adaptive advantages in frequently updating the information from other sectors of the environment—such as being able to react quickly to predators and prey, as a woodpecker on a tree trunk will continually interrupt its attention to food to observe the surrounding scene. And apparently limiting continuous attention to an upper limit of a second or so does not prevent animals and humans from performing operations of perception, memory storage, and problem-solving at adaptive levels. In any case, the eventual discovery of how attentional fluctuation is built into brain mechanisms should illuminate many of the questions raised here.

## Metaphors of Selective Attention

Most of the current metaphors of attention take into account only some of the six properties of the attended area just described. Most of the metaphors are based on familiar mechanical or physical devices like filters, spotlights, zoom lenses, gates, and channels. One problem with using devices like these to refer to attention is that properties of the device may be inappropriately included in the model along with features that do usefully describe how attention operates. Probably the clearest example of this unwarranted extension of a metaphor is the assumption that attention, because its boundaries may be considered similar to the edges of a spotlight, moves from one subject to another in the continuous way that a spotlight moves, which may not be the case at all.

I find it instructive to classify these metaphors according to whether they imply that attention is selected by enhancing the information flow in the target area or by inhibiting the information flow in the surround, or both. The metaphors that appear to select by facilitation of the target area while leaving the surround unchanged are: spotlight, zoom lens, resource allocation, and gain (increase in power). The metaphors that appear to select by inhibition of the surrounding area while leaving the target area unchanged are: filter, gate, and channel.

These seven metaphors, considered as mechanisms of attention, can be grouped under a general information-processing description when they are applied to the way they express attention in a parallel pathway of signals: greater information flow in pathways corresponding to the attended area relative to the information flow in pathways corresponding to the unattended surrounding areas. Apparently, no well-known physical device conveys selection by a combination of facilitation of the target information flow and inhibition of the flow in the surround. It is perhaps a fortunate coincidence that the term *selection* itself is unrelated to any particular physical device and therefore can remain neutral with respect to the choice of algorithm. Therefore it is relatively easy to consider the expression of attention separately from the particular mechanism that gives rise to that expression.

The foregoing discussion of the metaphors and properties of the selected area of attention illuminates the ways that notions of attentional mechanisms and expressions described at the cognitive level can be clarified and compared so that they may more effectively be evaluated in terms of their neurobiological plausibility. When the term *attentional expression* is used in the abstract computational sense, it refers to the output of the algorithm that runs on the mechanism of attention, whatever neuroanatomical mechanisms the nervous system may be said to use. This book is concerned with the way that attention operates in the human brain, however, and therefore serious consideration will be given only to metaphors that are not incompatible with the way that the human brain is known to operate.

## Selection of an Object in a Cluttered Field

Psychological studies of the properties of the selected area have uncovered several variables that influence the separation of target object information from distractor information. Three of the more effective factors are described here.

*Proximity of target and distractor objects.* In a seminal experiment (already mentioned above, in connection with attentional boundaries), Eriksen and Eriksen (1974) presented a horizontal array of letters containing a target letter at the center and asked subjects to identify that letter by moving a lever in one direction (for C or S) or another (for H or K). Positioned on both sides of the target letter were other letters chosen from this set of four letters and from a set of additional letters that were not assigned to either lever response. Although they were instructed not to attend to the letters on each side of the target letter, subjects showed clear evidence that they did at least briefly attend to the adjacent (distractor) letters, because the response time to a given target was increased as much as 80 msec when its distracting letters were those that would require moving the lever one way while the targeted letter required moving it in the opposite direction. This difference in response times to the incompatible and compatible target/distractor displays was labeled $I - C$ in the discussion above.

When the distracting letters were associated with neither response, the response times fell between those for the incompatible and compatible distractor displays, both for Eriksen and Eriksen (1974) and for LaBerge et al. (1991). The implications of this finding is that lateral masking of the target object by the adjacent distractors may lengthen the processing time needed for identifying the target in addition to the response-interference effect generated when the subject selectively attends to distractors.

Eriksen and Eriksen (1974) also varied the spacings between the letters from 0.06 degrees to 1 degree and found that, when target and adjacent letters were connected to different responses, the $I - C$ value decreased from 80 msec to 20 msec. Meanwhile the overall response time to different responses decreased from about 540 msec to 450 msec. A similar relationship between target-flanker spacing and response time was shown in another study (Eriksen and St. James, 1986) that displayed eight letters in a circle having a diameter of 1.5 degrees of visual angle. Therefore the proximity of the target and distractor apparently has a substantial effect on the time needed to process the target.

The effect of target-flanker spacing on accuracy of response was shown in an experiment by Estes (1982), who presented triplets of horizontally arranged letters at and away from the center of fixation under tachistoscopic viewing conditions. The more closely packed displays produced significantly higher errors.

*Similarity of target and distractor objects.* A second important variable affecting selection is the similarity of the target to objects in its immediate surround. Eriksen and his colleagues (Eriksen and Eriksen, 1974, 1979; Yeh and Eriksen, 1984) found that when target letters shared features with the distracting letters, the time needed to respond to the target increased. However, effects of target-distractor similarity on the accuracy measure under tachistoscopic exposures appear to be non-monotonic (Estes, 1982). Subjects made more errors as target-distractor similarity increased, but at the point of maximum similarity, when distractors were identical to the target, errors decreased relatively abruptly. LaBerge and Brown (1989) varied target-distractor similarity and found moderate increases in response time and small decreases in accuracy as the similarity between the target and distractor increased (see Figures 3.3 and 3.4). However, this experiment did not test the case in which distractors were identical to the target.

*Distance of target object from its anticipated location.* The third effective variable is the distance from the anticipated location that the target-distractor ensemble is presented. In one study (LaBerge and Brown, 1989), the effect of target-distractor similarity on response time increased the farther away the target appeared from the center fixation point, where subjects apparently had the greatest anticipation that it would appear. The data from the Estes (1982) tachistoscopic experiment showed a slight but consistent trend of the same kind for accuracy measures.

All three of the foregoing variables can be manipulated to change the difficulty of selective attentional processing in a visual task. Kahneman et al. (1983) referred to these kinds of effects as the "cost of filtering," in the context of the demand on attentional resources. In a series of experiments, he and his colleagues revealed subtle ways that target-distractor effects can be manipulated (Kahneman and Chajczyk, 1983; Kahneman and Henik, 1977).

Knowledge of effective ways to increase the amount of processing needed to separate information of a target from its surround can help investigators using physiological measures determine what areas of the brain are most active when selective attention is operating. A more difficult attention task will increase blood flow and glucose uptake in relevant brain areas, producing stronger PET effects (e.g., Corbetta et al., 1991; LaBerge and Buchsbaum, 1990). Similarly, the difficulty of a task can increase the rate of cell firings as recorded by microelectrodes (Spitzer et al., 1988). These measures will be described in more detail in later chapters that deal with neurobiological studies of attention.

## Experimental Tasks

*Tasks involving single operations of selective attention.* The ideal investigation of selective attention would allow only one attentional operation to occur each time a visual display is presented, so that the measure of processing reflects the operation of only that one operation. Most of the behavioral tasks described to this point are designed to direct attention to one object location or object attribute. But even for the relatively simple task of identifying a target item surrounded by distractors, knowing the exact location of the target does not guarantee that identification is achieved after one positioning of the selective area of attention. Experiments indicate that the effects of several attentional alignments are combined, and from these we must infer the characteristics of each component alignment of the attention area.

*Search Tasks.* In some experiments, the number of attentional operations provides a means of investigating particular properties of attention. Chief among these tasks are search tasks, in which many alignments of the attended area are made to objects in a display before a response is emitted. From a collection of response-time measures the experimenter attempts to determine how the operation of alignment was carried out on each object during the search.

Probably the most enduring principle to come out of a decade and a half of research concerns the conditions under which attention will and will not be required in finding the target object (Treisman and Sato, 1990). In the general case, when the target object differs from its surrounding objects with respect to a conjunction of separable features, then attention is aligned successively to each object in the display until the target is identified. When the target object differs from its surrounding objects with respect to one salient feature or texton (Julesz and Bergen, 1983), attention is not required because earlier processing mechanisms induce the target to "pop out" of the group of objects. Subsequent studies have suggested some modifications of the way the conjunction-attention principle is applied in certain types of displays. In the case of searching for a red vertical line among blue vertical lines and red horizontal lines, the number of items to be searched can be reduced by first inducing the red items to pop out (by priming the feature "red") and then serially aligning attention with each of the red items until the vertical line is identified (Egeth et al., 1984). In some displays, the subject may pre-select a subset of items on the basis of location (instead of color, as in the previous example) and then serially search these "clumps" (Pashler, 1987). In displays in which subjects are trained to prime two or three features of a target object simultaneously, the combined saliency value may be high enough to induce the target to pop out in the manner observed for targets defined by a single feature (Nakayama and Silverman, 1986; Wolfe et al., 1989).

When serial search of a display does take place, as evidenced by the positively sloped function relating response time to the number of elements in a display, the shifting of attentional alignments is estimated to take place at a rate as high as once every 80 msec (Treisman and Sato, 1990). When one serves as a subject in conjunctive search tasks, one gets the distinct impression that the examination of each item is taking place faster than one could voluntarily align attention to each item. It seems as though the rapid successive alignments of the attended area were under the control of an automatic mechanism, one that shifted attention following the relatively fast identification of an object's category as opposed to the slower identification of the location of the object itself (see, e.g., Kahneman and Treisman, 1992; Kanwisher, 1993). There appear to be obvious adaptive advantages in possessing a fast and automatic search mechanism. The moment during search when attention seems to operate most vividly and for a comparatively longer duration is that moment when the target is found. This relatively "late" manifestation of attention appears to correspond to the "decision-based" attention developed by Shaw (1978) and currently emphasized in the work of Palmer (1993; Palmer et al., 1994).

*Fluent reading.* The operation of attention during fluent reading resembles serial search with respect to the fast alignment of attention to visual items. In the case of reading, however, the location of the target is always known: it is either immediately to the right of the currently attended item or, when the reader reaches the end of a line, at the left end of the line of print below. The slow progress of the beginning reader is marked by the effort of voluntarily attending to each word, but fast, fluent reading gives the impression that words are processed automatically, without the participation of voluntary attention. When an unfamiliar word is encountered, however, or when a sentence does not seem to make sense, then words are examined one by one with voluntary attention. It is not being suggested here that the visual processing of words does not involve selective attention during fluent reading, when, it is assumed, one's voluntary attention is directed toward the intellectual domain where meanings of the words and phrases are being processed. Rather, as in the case of fast visual search, the attentional alignment to each visual item has come under the control of some mechanism other than voluntary attention. This state of affairs is deemed beneficial, because it frees voluntary attention to participate more fully in semantic processing (see LaBerge and Samuels, 1974), where it is assumed that attention is expressed in terms of selected enhancements of processing but in a space in which objects are coded in a more highly distributed manner than in most perceptual spaces.

The time taken to select an object in reading and search tasks is strongly influenced by two of the three factors described previously, that is, the proximity and similarity of distracting objects. Typical pages of text, like this one, pack words close together, both horizontally and vertically, and word shapes can be regarded as quite similar, owing to their sharing of letters and the relatively small number of line features shared by letters. Search displays vary more than textual displays in the arrangements of the objects and in their similarity to each other. Groups of cars in parking lots, crowds of people, arrays of objects in store windows, stacks of books on shelves, collections of pottery in corner cupboards—these are examples of displays that are frequently searched to locate a particular target object. But in all the examples of reading and search displays, distractors are present and increase the time needed to identify each object. Therefore, when these kinds of tasks are used in an experiment to estimate the time taken to shift visual attention from one location to another (e.g., Treisman and Gelade, 1980), or when displays containing no distractors are compared with search tasks (e.g., Julesz, 1991), the interfering presence of distractors, particularly in their proximity and similarity to the target, must be taken into consideration.

## Selective Attention to Actions

By far the majority of experimental and theoretical investigations of selective attention have been based on perceptual processing. Recently, however, an increasing number of studies have probed the effects of attention on processes that involve actions taken by an individual, particularly internal actions such as organizational or "executive" functions (e.g., Allport, 1989; Duncan, 1994; LaBerge, 1990; Posner and Peterson, 1990) that are involved in solving a problem. Examples are planning a way to hold two traveling bags while presenting an airline ticket to the attendant as you board a plane or planning the way to sequence a pair of phrases in a sentence you are about to utter or write down.

*Organizational anchoring by attention to goal images.* Many complex problems, such as those in intelligence tests, may be analyzed into subproblems, and for each subproblem there exists a solution that can be conceptualized or imaged as a goal. Often a subproblem may be analyzed into its own subproblems, and so on, so that the entire problem may be conceptualized or imaged as a hierarchy of goals (Duncan, 1994). As the individual solves the same problem repeatedly, each remembered subgoal may automatically signal the particular actions or operations that have achieved it. When an individual is engaged effectively in solving complex problems, attention is assumed to be involved selectively in at least three component operations: (1) selecting the appropriate subgoal from an array of alternative goals, (2) selecting the operation or operations associated with the achievement of that subgoal, and (3) sustaining attention to each operation in order to insulate it from interfering cross-talk from other concurrent operations. In the case of presenting an airline ticket to an attendant while carrying two traveling bags, a subgoal might be represented by an image, held in working memory, of holding the heavy bag in the left hand while the right hand holds the briefcase with three fingers and the ticket between the thumb and first finger. The subsequent goal is to release the ticket to the attendant without dropping a bag, which requires that attention be carefully constrained to the releasing of the ticket. Thus, if attention is directed toward hurrying to board the plane, then less attention is given to the selective release of the ticket, and the likelihood of a confused response becomes greater.

The expression of attention to subgoals and operations during complex problem-solving may be modeled on many of the attentional metaphors described for cases of making perceptual judgments. In the case of attention to the planning and execution of actions, both internal and external, it seems appropriate to gather the many metaphors of attention under a general information-processing description of their outputs as attentional expressions. A description of this sort, as noted above, focuses on a relative enhancement of information flow in pathways corresponding to the attended goal or operation relative to the information flow in the pathways corresponding to the unattended goals or operations.

## Summary

This chapter has examined one manifestation of attention, the selective processing of information. Most of the analysis and data concerned involvement of attention in perceptual processes, and a brief description was also given of selective attention to internal actions during problem-solving, a topic of increasing interest to investigators of attention.

After a brief introductory description of some of the historical roots of selective attention, six properties of the attended area, produced from the studies of many visual-attention researchers, were analyzed. Many of the metaphors of attention that guide much of psychological research may be compared with respect to the ways that they explain some of these properties. If the many metaphors of attention can be clarified in this way, they may more effectively be evaluated on the basis of psychological and neurobiological data.

# 3

## Preparatory Attention and Maintenance Attention

At any moment of waking life, some part or sector of the vast array of sensory and ideational information available to the individual appears to be emphasized over other areas of information flow. The many properties attributed to this selected sector of information flow, the "attended area," were described in the previous chapter. Although visual perception is usually used to illustrate these properties, it is commonly assumed that they would also be exhibited in other domains such as audition and action.

Attention may be sustained quite briefly, as when a visual display is scanned or a familiar word is read. But often the attended area is sustained for relatively long periods, especially when the location or attribute of an item is known in advance. When a driver, for example, expects to see a green light appear at a street intersection, the location of the light and the green color become the attended areas of information in spatial and color maps of the brain. The informational emphases that define these attended areas may be sustained without interruption over the few seconds between the appearance of a yellow light signaling cross traffic and the onset of the green light in the driver's lane. The few seconds of sustained attention in preparation for seeing the green light are quite long compared with the 50 to 100 msec estimated to be required for selectively identifying the light when it occurs or examining an object during a visual search task. Thus, selective attention apparently can operate at snapshot durations while preparatory attention usually operates on a scale of seconds.

It would seem, however, that preparatory attention to an expected event almost always involves selective attention. Attentional preparation to perceive the location and color of the expected green light is presumed not only to emphasize information flow in the visual modality but to focus in particular on a constrained location and color site in the appropriate visual maps of the brain. Moreover, the extended durations of preparatory attention can provide the time required to intensify attentional activity in those selected areas of processing to very high levels. At such times a racing driver may report that he or she was not just "mindful" of the impending green light—the image of the green light seemed to completely "fill the mind." As a consequence of this capability of sustained preparatory attention to produce high elevations of brain activity for relatively extended durations, tasks requiring prolonged attention have been incorporated in the the design of many brain-imaging experiments.

Selective and preparatory attention are treated separately here not because of a difference in the way that they are expressed, since both manifestations are assumed to arise from the same enhancements of information flow in particular sectors of brain regions. Rather, the difference stems from the fact that preparatory attention requires an expectation: it occurs before some expected perception or action and it can be relatively long in duration. Selection, on the other hand, can occur after as well as before any event occurs, expected or unexpected, and when it occurs after the event, its duration is typically quite brief as it shifts from one aspect of the event to another.

Expectation also distinguishes preparation from the third manifestation of attention: maintenance. Maintenance is similar to preparatory attention in that it too is typically sustained for longer periods of time than selection operates, but it lacks the immediate goal that expectancy lends to preparation. Maintenance attention has been researched less than the other two manifestations, and it is treated only briefly at the end of this chapter.

## Preparatory Attention

One important benefit of preparatory attention, as indicated in the first chapter, is that the individual can react more quickly to an expected event when it occurs. This increase in processing speed is assumed to be achieved by preprocessing in either perceptual regions or action regions of the brain, or both. One purpose of this chapter is to attempt to articulate what is involved in this kind of "preprocessing."

*Early descriptions of preparatory attention.* A backward glance through history to locate the earliest influential description of attentional preparation again brings us to the writings of William James. In his chapter on attention he classified the two "physiological" processes of attention as the accommodation or adjustment of the sensory organs and the anticipatory preparation from within of the ideational centers concerned with the object to which the attention is paid (James, 1890, vol. 1, p. 434).

James credited Wundt for the reaction-time experiments that supported the claim that attention shortens perception time. An example of Wundt's experiments in this area is one in which strong and weak sounds were presented to a subject under two conditions: in an alternating sequence, in which successful prediction of the stimulus was possible, and in an irregular manner, in which successful prediction was not possible. Wundt found that, when given in alternation, the strong and weak sounds elicited response times of 116 and 127 msec, respectively, but when the sounds were produced irregularly, the response times were 189 and 298 msec, respectively. The substantially shorter response times for the predictable intensities were viewed as indicating that the particular intensity had been anticipated. The results of such studies as these led James to conclude that "concentrated attention accelerates perception, so, conversely perception of a stimulus is retarded by anything which either baffles or distracts the attention with which we await it" (p. 429).

In these simple tasks, reaction time could also be shortened by anticipation of the response to be made. Lange (described in James, 1890, vol. 1, p. 92), a student in Wundt's laboratory, had contrasted "extreme sensorial" attention with "extreme muscular" attention, and found about a 100 msec advantage favoring the processing of the muscular task. While the amount of difference may be disputed, the important outcome of these experiments was that they suggested that anticipatory attention to motor processes could shorten the time involved in performing a response.

To describe what was going on during the time period when a person was anticipating a stimulus or a response, James claimed that the person was simply imagining the appearance of the stimulus or the muscular impressions of the response: "preparation . . . always partly consists of the creation of an imaginary duplicate of the object in the mind" (p. 439). James regarded the image of an object as a "pre-perception" of the stimulus, which in effect did part of the job of processing a sensory stimulus and thereby reduced the additional sensory processing required to trigger the perceptual event when the actual stimulus occurred. The "prior image" could be viewed as the preprocessing of a location of an object as well as the preprocessing of the attributes of an object; his ideas here anticipated some of the aspects of image processing currently being extensively investigated by Kosslyn (1988, 1994).

The prior image is presumably maintained by a memory in a higher-order system that continually refreshes the perceptual processor as the person waits for the stimulus to occur. Presumably, when the stimulus appears, the projected information from the higher-order system need not shut down but may continue to add to the stimulus input, driving the activity within the particular cortical area or areas to a higher level than would be produced by the stimulus alone. Thus, James conjectured, attending to a perceived stimulus consists of "a brain-cell played upon from two directions. Whilst the object excites it from without, other brain-cells . . . arouse it from within" (James, 1890, vol. 1, p. 441).

James's conjecture of a two-way activation of brain units, in modern terms from bottom up and from top down at the same time, apparently is holding up well in the face of data obtained with today's more sophisticated physiological measures. For example, PET experiments show that many brain areas activated by image-generating tasks are the same as those activated by the corresponding perceptual tasks (Kosslyn et al., 1993). Also, PET experiments use matching-from-sample tasks in which the sample serves to induce a preparatory state (Corbetta et al., 1991) in anticipation of the upcoming target. When a specific attribute of the target stimulus (e.g., its color, shape, or movement velocity) is cued by the sample, specific brain areas corresponding to these attributes show more activation than they do when no specific attribute is cued. This finding suggests that cueing of an attribute induces top-down preparatory activity in the particular circuits that specialize in the bottom-up sensory processing of that attribute. In addition, ERP experiments (e.g.,

Mangun et al., 1992) showed increased amplitude of wave-form components as early as 100 msec after stimulus onset, when subjects were cued to expect a target in a particular location. Also, single-cell recordings (e.g., Spitzer et al., 1988) show increased firing rates in color- and orientation-sensitive cells in matching-to-sample tasks when a particular color or orientation is cued prior to the onset of the target.

Sixty years after James, Hebb (1949) articulated a view close to his in terms of the activity in cell assemblies. "Each assembly action may be aroused by the preceding assembly, by a sensory event, or—normally—by both." The central facilitation from one of these activities to the next was regarded by Hebb as the prototype of "attention." More recent indication of research interest in the preparatory manifestation of attention was the collection of papers in the book entitled *Preparatory States and Processes* (Kornblum and Requin, 1984).

## Attentional Preparation or Cognitive Expectation?

Often the terms *preparation* and *expectation* are used interchangeably, a habit that confuses an important distinction, particularly when one is describing attentional phenomena. A preparation for a particular stimulus or action is an elevation of activity in the corresponding perceptual or action brain area that speeds processing of stimuli or actions when the appropriate triggering event occurs. An expectation is an item stored in long-term or working memory that codes an event in terms of its attributes and its spatial and temporal characteristics. This code may be in a verbal form, which means that the existence of a expectancy in long-term or working memory need not produce an elevation of activity in brain areas that process the perceptions and/or actions that correspond to the stored expectancy. As words are read their meanings are often accessed into working memory without the corresponding perceptual or action images being generated as well. For many people, generating images of an object may require more effort than simply maintaining the verbal label for the object in working memory. For example, a verbal consideration that a long-lost pet might come home does not begin to "fill the mind" as does attending to the image of a golden retriever poking its head through the doorway.

The term *priming* is customarily used to describe the activation of perceptual and/or action processes, and whether priming may be induced in a person is, in many cases, under the voluntary option of the subject. For example, while waiting for the light to change at an intersection, a driver may be told that the light is about to change but may not prime perception of the location and color of the light by attending to the specific location and imagined color, even though the message is held verbally in working memory. Instead, she may opt simply to increase readiness to respond by moving her foot toward the accelerator. Thus, as viewed here, preparations to perceive objects and preparations to perform internal and external actions are assumed to be expressions of attention, whereas expectations of occurrences of particular objects or actions are assumed to be expressions of memory. Expectations often exist without the accompanying preparations, but it is less conceivable that preparations could exist without expectations. Sustained attention can take place without expectations, however, but this case is viewed as maintenance attention and will be discussed in the chapter that follows.

A few examples may serve to clarify this distinction as it is met in daily life. You walk to your customary parking lot and expect to find your car in a particular location, and if the car were not there you would experience surprise. The expectation of finding the car may or may not be accompanied by the attentional state of preparing to perceive its characteristic shape and color, so that as you approach the parking lot your attention may be intensely directed toward solving an intellectual problem or, if it is raining, your attention may be concentrated on staying dry under an umbrella. Or, while listening to familiar music you are surprised by a wrong note, indicating that an expectation had been violated (Carlsen, 1981; Unyk and Carlsen, 1987), but at the time the wrong note occurred your attention may have been directed to the program notes. Generally, at the ending of a piece a listener's expectancy of the sequence of the last few sounds is regarded as sufficiently important to induce attentional preparation for each sound just prior to its occurrence. Indeed, it has long been a practice by performers to retard the tempo at the ending of a piece in order to involve the attention of the listeners, many of whom have let their attention wander to something else during the course of the music. Carefully timed retards of ending cadences can synchronize the preparatory attention of an audience so precisely that when the last sound ends, the audience initiates their applause virtually simultaneously.

While the foregoing examples illustrate differences between expectations and attentional preparations for perceiving sensory events, other examples illustrate differences with respect to performing actions. You may, for instance, sometimes drop a spoon while you are washing dishes. The strong expectation that you have a sure grasp on the spoon during its transit from the tub of water to the drying tray had been set up from extended practice; the accidental release of the spoon therefore produced a surprise. Evidently, the particular action (or lack of it) that induced the dropping of the spoon was not the action expected. The preparatory attention for the response that would have ensured the appropriate action and prevented the "accident" was directed elsewhere at the time.

Expectations and preparations can also be distinguished in the sequence of responses involved in writing or typing text, to take another example. We expect that familiar words will be spelled correctly as we write them, but on occasion spelling errors appear. We experience a mild surprise when we realize the error, which is an indicator that an expectation had been thwarted. When we correct the spelling error, we typically attend more closely to the actions involved in writing, anticipating each letter just prior to our writing or typing it. One could say that in the normal writing of text, our attention is directed to meanings instead of to words or letters, and although we do not specifically anticipate or prepare the execution of each letter, we do maintain expectations of the appropriate letter sequence of the words we write.

The distinction between attention and preparation may also be seen in musical performance. Performers who prepare a piece by repeatedly practicing it in the same form often find their attention wandering while they play it, because the sequence of actions has become highly automatic. Thus, the sequence of responses is run off in accordance with a particular motor expectation but with little attentional preparation or anticipation. Consequently, when performing in public, one's attention is free to be directed to other events, such as the presence of the audience, which may bring to mind the social consequences of making a mistake. Attention to the costs of making a mistake leads to a condition described as "performance anxiety." On the other hand, a musician who practices by continually anticipating the musical content of upcoming notes and phrases (practicing "in the music" versus "in the technique"), as one almost always does when one improvises, is more likely to attend closely to the music in performance situations and not be distracted by the prospect of mistakes. An occasional mistake that does not detract from the "music" is typically deemed less important when both performer and audience are attending to the musical content rather than to technical execution.

For a final example, consider what is implied by "getting" a joke. Responding with laughter often requires that the listener anticipate the events of the joke as they unfold so that when an unexpected event occurs (delivered in the punch line), it produces an intense surprise effect. Skillful joke-tellers choose their wording and adjust the timing of their delivery so that when the punch line occurs, their listeners have focused attention strongly in anticipation of events or classes of events other than the one that the punch line is based on. If, on the other hand, listeners simply follow the delivery without anticipating the ending, they may experience some surprise, owing to the thwarting of the expectancy, but less than if they had actively anticipated events as the joke was told.

Preparatory attention is frequently employed when the "expected" event does not occur but could occur, particularly when the "expected" event is a mishap. A waiter carrying a tray of wine glasses down a flight of stairs is usually compelled to prepare to see them tip to one side or another. Preparatory "pre-processing" of a slight tipping of the glasses enables him to make a fast detection of the slightest deviation from the vertical and then a successful correction. Unless attention is voluntarily sustained in cases like this, a mishap is likely.

In general, most people routinely anticipate pleasures and displeasures during the course of their daily activities. While driving to the beach on a sunny day, for example, you expect to enjoy the surf and sun, and this expectation will be confirmed or disconfirmed once you have spent some time there. But during the drive you may, in addition, actually anticipate the warm feeling of the sun bathing on your face and the smell of suntan lotion. Clearly, the involvement of attention is much greater when you anticipate through sensory imaging than when you simply acknowledge verbally that you expect to enjoy the sand and sun at the beach. Furthermore, anticipating pleasures through imaging would seem to produce more intense pleasure than sustaining the expectation of pleasure (in working memory) while attending to momentary activities, such as driving a car or viewing the passing scene along the road. Indeed, the hours that people accumulate in imaginative anticipation of a pleasurable event may exceed the hours spent in experiencing the event when it actually occurs. Hence, it is not surprising to hear some people say they get more out of the anticipations than the realizations of events such as parties, vacations, and the like.

Whether or not they are accompanied by preparatory attention, expectations require the storage of a representation of some object or event that will occur at a particular time, usually in the near future. This object/time representation is assumed to be located in a working-memory type of storage, where it is assumed to be influenced by self-induced instructions or by instructions given by others. If attention is engaged by the object/time goal, higher-order processes produce activity in sensory domains appropriate to the object, which constitutes the expression of preparatory attention.

*Experimental methods of studying attentional preparation.* Historically, preparatory attention has been investigated mainly with two types of tasks, the delayed-reaction task and the vigilance task. In the delayed-reaction task the subject's attention must span delays between the cue stimulus and the reward stimulus in which no external event sustains information. For example, a subject (e.g., a monkey or child) is shown two objects, and one object is briefly lifted to reveal a reward underneath it. Then a shield of some kind is interposed between the subject and the display for a particular length of time. After this delay period, the shield is withdrawn and the subject lifts the object that conceals the reward. Two types of object anticipations are customarily used in delayed-reaction tasks: anticipation of object location, in which the object shapes are the same, and anticipation of object shape, in which the objects are of different shape and their locations are varied. In location-anticipation tasks, the subject can maintain a bodily posture during the delay that is directed toward the location of an expected goal object, thus carrying information about the object's location by a response anticipation. If experimenters wish to induce subjects to code the location of a goal object in a representational manner, they must devise a means to prevent the animal or child from maintaining these informative postures during the delay interval (Diamond and Goldman-Rakic, 1989).

Most delay tasks are designed so that postural indicators of the location of a stimulus shape is unproductive for the subject, and therefore the time interval must be spanned by some representation stored in working memory. Of course, a subject can be trained to associate a particular object with reward by extended practice, so that the delay between the first and second presentations of the goal object presents no problem. In this case, it could be said that the subject expects or even anticipates the object that has consistently been associated with the reward. But for experiments aimed at demonstrating that working memory spans the delay period, performance of the task by means of associative learning must be prevented, usually by varying the relationship between stimulus objects and rewards from trial to trial. When this is done one begins to see relatively sharp differences in both the animal phylum and ages of children in the successful performance of the task (Diamond and Goldman-Rakic, 1989).

Expectations are assumed here to be states in working memory, and long-term memory of an item can refresh the working memory from time to time to sustain long durations of an expectation. One task that exemplifies very long term extensions of expectancies are vigilance tasks (e.g., Mackworth, 1969). When a sailor scans the horizon for objects during the long night watch, his perceptual processing is guided by sustained expectations of objects possessing particular attributes and appearing in a particular range of locations. The set of expected goal objects and events may be relatively specific and the expectation, though of a low-probability event, may be sustained over relatively long periods. Preparatory attentional states, on the other hand, cannot be sustained over long periods of time, particularly if the states involve high levels of attention to detect low-luminance targets, as laboratory investigators are well aware. One reason is that voluntary preparatory attention is marked by relatively rapid fluctuation within a period of a few seconds, while the memory of an expectation can be continually refreshed by the environmental context and remain comparatively stable over long periods of time. Vigilance could be characterized as an extended state of expectancy that is peppered with occasional and brief periods of preparatory attention.

The consistent decrements in performance over time shown in vigilance experiments (Broadbent, 1971; Mackworth, 1969) may be influenced by an apparent decrease in alertness, particularly when target signals are rare and brief. One of the important variables that influence the alert state is the level of the neurotransmitter norepinephrine in cortical pathways, which appears to increase the signal-to-noise ratio in neural signaling (for a review, see Robbins and Everitt, 1994). PET studies have targeted the right frontal and parietal areas as particularly active during sustained vigilance in humans (Pardo et al., 1990). Posner (1994) has integrated these findings into a network model of alerting, particularly as it applies to brain areas serving visual attention.

In the context of the present characterization of attention, the level of norepinephrine in cortical pathways may set limits on the difference between activity at the attended sites and the activity at surrounding sites. At low norepinephrine levels in which baseline noise is high, it may be difficult to produce relatively high activity in an attended area. In contrast, at high norepinephrine levels in which the noise level is low, it may be considerably easier to obtain relatively high activity in an attended area.

To summarize the foregoing description of the expectation/preparation distinction: A main goal of attentional preparation of an event prior to its occurrence is the more effective processing of information arising when that event eventually occurs. But to maintain active attentional preparation of an event over time requires that a representation of that target event be stored in working memory during the time interval of preparation, which the subject voluntarily decides to use to sustain an appropriately high level of activity in the sensory or action systems corresponding to that representation. For example, anticipation of a visitor's knock on the door in the next seconds following the sound of footsteps involves a representation of knocking in working memory that is time-tagged in the immediate future. This representation in turn may be used to engender activity in the auditory cortex that pre-activates auditory pathways corresponding approximately to the auditory features of the expected knock stimulus. Then, when the knock actually occurs, it will be processed more rapidly because the auditory detection process has been advanced through the pre-activation process of preparatory attention. In short, the *expression* of preparatory attention is the elevated activity in a selected sector of a sensory pathway of the system; a *mechanism* of attention can selectively enhance the level of activation corresponding to the degree of attentional concentration, and the *control* of the expression and mechanism in preparatory attention is assumed to arise mainly from a corresponding expectancy in working memory.

## Perceptual Preparations for Objects and Their Attributes

In the 1960s there began an active development of experimental procedures that were aimed at inducing specific preparation in the simple type of reaction-time task used by psychologists near the turn of the century (e.g., by Wundt, Exner, Cattell, as described in James, 1890, vol. 1, chap. 11). In the simple choice task, in which one response is assigned to one stimulus (such as a red square) and a second response assigned to the other stimulus (a blue square), the subject can be induced to respond faster to a particular stimulus simply by presenting that stimulus more frequently than the other stimulus. However, it is not clear whether the faster response time is due to anticipation of the stimulus, the response, or both. In an attempt to separate stimulus and response anticipations, one can assign two stimuli (red and blue) to one response (a left button) and another stimulus (green) to the other response (a right button), and then vary the relative frequencies of the first two stimuli. Differences in mean response times to the two stimuli (red and blue) assigned to the same response would then presumably reflect differences in anticipations of the stimuli and not the response, since the response was the same in both cases.

In one such experiment (LaBerge and Tweedy, 1961), the relative frequency of the two stimuli assigned to the same response were 0.50 and 0.10 (and 0.40 for the stimulus assigned to the other response), and the mean response time to the more frequent stimulus was 30 msec less than that given to the less frequent stimulus. At about the same time, another experiment (Bertelsen and Tisseyre, 1966) assigned two stimuli to one response and two to the other response, and when the relative frequencies of the two stimuli assigned to the same response were 0.70 and 0.30, the mean response time to the more frequent stimulus was 100 msec less than that given to the less frequent stimulus.

In a subsequent experiment (LaBerge et al., 1967) that used the same stimulus-response assignments, it was shown that a 120 msec stimulus-anticipation effect could be induced by offering a greater reward for fast responses to one stimulus than to the other stimulus while maintaining equal frequencies of presentation of the two stimuli. The data from these three experiments suggested that response time could be reduced by anticipating the stimulus without varying anticipation of the response.

With the same type of experimental design, one can attempt to answer a complementary question: can response time be reduced by anticipating the response without varying anticipation of the stimulus? One such experiment (LaBerge, LeGrand, and Hobbie, 1969) showed that the mean difference in response time varied in a continuous manner as the relative frequencies of the two responses varied from 1:1 to 9:1. Furthermore, when relative frequency of both stimulus and response were systematically varied, the resulting differences in mean response time varied accordingly. Thus, in a relatively simple choice task, it appeared that anticipations for stimulus and responses could be varied independently and that they could occur together, at least within the same block of trials.

One of the problems with this conclusion is that the measure used average response times over trials, and therefore trial-to-trial anticipations could vary in a wide variety of ways, including the case in which anticipation assumed either zero or one particular positive value. For example, in the two-stimulus, one-response case, the subject may anticipate maximally the frequent stimulus on a high proportion of the trials and anticipate maximally the less frequent stimulus on a low proportion of the trials, with the proportions perhaps following the probability matching law (Estes and Straughan, 1954). The problem seems to stem from the fact that when anticipation is induced by raising the probability of a stimulus, the probability of a stimulus is defined across trials, not within a trial. Similarly, when reward is given on the basis of mean response times, the mean is defined across trials, not within a trial.

What was needed in these kinds of experiments was to introduce a preparation-controlling event within a trial instead of across trials. One technique (LaBerge, Van Gelder, and Yellott, 1970) presented a cue just prior to the target stimulus on each trial that predicted the target stimulus that would appear on that trial. Four target stimuli were assigned to two responses, for example, red and high tone were assigned to a left button response, and green and low tone were assigned to a right button response. On each trial, one of these stimuli was given as a prime for 200 msec, followed by a 3,000 msec interval in which the appropriate anticipation could build before the target stimulus appeared. The target remained on until the subject responded. For example, a trial began with a 200 msec flash of a red square, and after 3,000 msec the red square would appear and the subject responded with the left button press. On some percentages of the trials, one of the other stimuli or a blank occurred, but on 73 percent of the non-blank trials the red square target followed the red square cue.

The mean response time for validly cued trials was approximately 270 msec, and for invalidly cued trials it was approximately 380 msec. The 270 msec response time could be lowered by only 10 msec when the validity of the cue was increased from 0.73 to 1.0 (on non-blank trials). Moreover, analysis of response times as a function of runs of a particular target stimulus showed no change in mean response time, indicating that subjects were treating trials independently. Analysis of runs in the LaBerge and Tweedy (1964) study, which had varied probabilities of targets across trials, had shown very strong run effects: as the run length of one stimulus increased, the response times to that stimulus in the next trial increased while the response times to the other stimulus decreased. (For a more detailed analysis of the response-time run effects in choice experiments, see Laming, 1968). Therefore we concluded that a predictive strength of 0.73 for a cue-same-as-target trial induced a subject almost always to anticipate the target indicated by the cue.

One other methodological change was introduced in an attempt to control more effectively the anticipation processing on each trial. Immediately following a response, a subject was given a low-amplitude burst of white noise if the subject's response time was less than a criterion value. The criterion value was calculated by adding 20 msec to the mean response time of the previous block of trials.

The results of this experiment not only provided evidence favoring consistent trial-by-trial anticipations of a particular color or tone, but also furnished estimates of the time required to shift attention from a particular stimulus to another, when the shift was either within a modality (as from red to green) or across modalities (from red to high tone). Estimating cost and benefit (Posner and Snyder, 1975) separately requires an estimate of response times to a neutral stimulus, which in this experiment was provided in a separate condition that presented the same cues, each of which predicted the four target stimuli with equal probabilities. The resulting cost/benefit values of 40/70 msec are not far from the cost/benefit values of 36/85 obtained by Posner and Snyder (1975), who used a cue-validity probability of 0.80 with letter stimuli.

*Imagery and preparatory attention.* The assumed relationship between visual priming and preparatory attention discussed in the foregoing section may be extended to the way imaging has been related to anticipatory or preparatory attention (e.g., Farah, 1989; Kosslyn, 1994; Intons-Petersen and Roskos-Ewoldson, 1989; Rosch, 1976). In particular, Kosslyn has taken as his working assumption that the generation, maintenance, and use of visual images involves the same brain mechanisms that underlie visual perception itself. Data from many cognitive experiments appear to support this principle (e.g., Farah, 1989; but objections do exist: see Heil, Rosler, and Hennighausen, 1993), and recently PET studies (Kosslyn et al., 1993) have shown that visual perception and visual imaging tasks activate many of the same cortical areas, particularly posterior cortical areas known from other studies to be crucial to early visual processing of objects and prefrontal areas known to be crucial to top-down control of early processing. The general conclusion from these imaging studies appears to be that instructions, coded in working memory, can generate activity in the same visual systems that process sensory input. The relevance for preparatory attention is that expectancies stored in working memory can be combined with imagery instructions to elevate activity in perceptual circuits corresponding to the shape and/or attributes of the expected object, which is the expression of preparatory attention.

## Perceptual Preparation for Locations of Objects

In a series of experiments in the early 1970s, C. W. Eriksen in the United States and A. Van der Heijden in Europe began their extensive series of experiments of the attentional effects of spatial cues in visual displays (for a review, see Van der Heijden, 1992). Van der Heijden and Eerland (1973) showed increased detection accuracy of the letter O when foreknowledge was given of its location, anticipating the later finding by Bachinski and Bachrach (1980). Eriksen developed the display in which stimuli are presented at a point or points along a clock-like circle (Eriksen and Hoffman, 1972). A typical cue was a bar marker presented somewhere on the clock-circle 50 to 100 msec before the target letter appeared. They found that the advantage in response time to the target of a marker cue over no cue was 30 to 40 msec. Similar response-time advantages of foreknowledge were obtained by Posner et al. (1978) in a spatial task that required the subject only to detect the appearance of an asterisk in one of two boxes located on each side of a central fixation cross. The probable location ($p = 0.80$) of the asterisk was marked by the illumination of the box outline in which it was to appear. Response times to the anticipated target location were 35 msec faster when a cue was given, within the range of values revealed in Eriksen's circular displays. Also noteworthy in the Posner experiment is the response time in those 20 percent of cases when the cue and the asterisk appeared in different locations: detection took 35 msec longer after a misleading cue was given than when no cue was given. More recent experiments have confirmed the general finding that cuing the location of an object before it appears can benefit its processing after it appears (Downing, 1988; Hawkins et al., 1990; Müller and Findlay, 1987; Müller and Rabbitt, 1989; Sagi and Julesz, 1986).

*Orienting.* The early experiments by Posner and his associates led to a very influential theory of attention based on brain networks involved in visual orienting. In its most recent form (Posner, 1994), Posner's general theory of attention contains three networks: (1) an orienting network that operates during early visual processing in posterior cortex, (2) an executive network in the mid-frontal lobe that operates in the detection of events and the voluntary control of orienting on the basis of current goals, and (3) an alerting network, referred to above, that appears to involve amounts of the neurotransmitter norepinephrine in cortical pathways, particularly in the right frontal and parietal lobes. The present discussion is particularly concerned with the relationship of orienting and the orienting network to preparatory attention, since orienting cues presented prior to a stimulus typically engender this type of attentional manifestation.

The familiar definitions of *orient* include "to adjust or bring something into due relation to its surroundings, circumstances, or facts," and "to direct or position something toward a particular object." In the context of visual spatial attention, this sense of adjustment emphasizes the change of attentional alignment from a current object location to a new object location. Although orienting had been traditionally studied as a reflex-like response (Pavlov, 1927; Sokolov, 1960), Posner has used the term *orienting* to denote "aligning attention with a source of sensory input or internal semantic structure stored in memory" (Posner, 1980). Posner distinguishes three neurally based components of the change in attentional alignment: disengagement from the present location, movement to the new location, and engagement at the new location (Posner, 1994). Operations that accomplish or produce disengagement, movement, and engagement are assumed to be occur in the parietal lobe, midbrain (superior colliculus and its adjacent areas), and the pulvinar nucleus of the thalamus. The movement of attentional focus can take place without movement of the eyes, but because the inferred movement of attention resembles in some respects the way the eyes overtly orient by moving in their orbits attentional movements are often referred to as "covert orienting."

The experimental procedure that has been used extensively within this orienting context appears to induce a form of preparatory attention in subjects that enables them to respond more rapidly to a target stimulus in a new location. The standard display contains two unfilled boxes (about 1 degree in size) that are positioned approximately 8 degrees to the left or right of center. One of the boxes is briefly illuminated, and after a variable delay an asterisk appears in that box or the other box. The box that is illuminated typically shows the asterisk (the target) on 80 percent of the trials (valid cued trials), while the other box shows the asterisk 20 percent of the trials (invalid cued trials). As noted above, responses to the target are more rapid when preceded by a valid cue (Posner et al., 1978), and the target can be more accurately detected at low illuminations when it is preceded by a valid cue (Bachinski and Bachrach, 1980; Downing, 1988). In addition, scalp electrodes reveal elevated electrical activity in brain areas corresponding to the location of the valid cue (Mangun and Hillyard, 1987). These benefits in processing the target occur within the first 50–150 msec after the onset of the valid cue.

In this standard cued-location experiment the measurements apparently most directly indicate the processing that takes place during the alignment of attention at the new location, and less directly they indicate the processing that takes place during disengagement and the movement stages of orienting. This emphasis on attentional alignment at the new location is consistent with the way that Posner has used the term *orienting* (Posner, 1980). An interpretation of the facilitation of processing at the target location favored in this book is that the cue increases activity in the corresponding brain maps so that the target that follows will be processed more rapidly and accurately. Viewed in this manner, the cue could be said to generate preparatory attention at the target location prior to the onset of the target.

The cued-location task has been used by Posner and his associates to investigate more closely the operations of movement (Rafal et al., 1988) and disengagement (Posner et al., 1984), as well as the engagement operation, in patients with lesions in the midbrain and parietal lobe (detailed discussion of this work is postponed to Chapter 6). Others have investigated the dynamic aspects of orienting corresponding to the disengagement and movement stages using somewhat different ways to cue an object's location.

Jonides and Yantis (Jonides and Yantis, 1988; Yantis, 1993) and others (e.g., Bacon and Egeth, 1994; Henderson and MacQuistan, 1993; Klein and Hanson, 1990; Kingstone, 1992; Koshino et al., 1992; Theeuwes, 1992, 1992) have explored the conditions under which a cue induces orienting in visual space. (For related studies of location indexing by multiple cues, see Pylyshyn et al., 1994; Trick and Pylyshyn, 1993; Wright, 1994). There are two main ways of cuing an object's location: an *exogenous* cue initiates orienting from the onset of a stimulus away from the current alignment of attention; an *endogenous* cue induces higher-order processes of the subject to initiate orienting. An example of an exogenous cue is the sudden brightening of one of the boxes in the cued-location task used by Posner; an example of an endogenous cue is the presentation of an arrow at the center of a display that signals the location of the target. While the top-down control of orienting induced by an endogenous cue typically involves voluntary processing, the bottom-up control of orienting initiated by an exogenous cue is typically similar to a reflex, and for this reason it resembles the older notions of orienting described by Pavlov (1927) and Sokolov (1960).

The time taken to orient following an endogenous cue is almost always more than the time taken to orient following an exogenous cue. This difference led Nakayama and MacKeben (1989) to label the two types of orienting "sustained" and "transient," suggesting an analogy with sustained and transient cells in the retino-geniculate pathway of early visual processing. The voluntary versus automatic contrast between endogenously and exogenously controlled orienting has been emphasized by several investigators (e.g., Müller and Rabbitt, 1989; Weichselgartner and Sperling, 1987).

When an exogenous cue consists of an intense, unexpected stimulus and the current attentional concentration is not strong, then it is highly likely that the cue will initiate orienting to its location (see Yantis, 1993, for a review of relevant experiments). The sudden onset of an audible tone, for example, produces a much greater amplitude in the N100 wave component (the electrical effect recorded from the scalp 100 msec after the onset of a stimulus) when it is the first tone of a sequence of tones that follows an interval of silence of several seconds (e.g., Fruhstorfer et al., 1970). The orienting response is known to be accompanied by a broad spectrum of effects, including an arrest of ongoing behavior, a delay in breathing, and a deceleration in heart rate (Lynn, 1966; Naatanen, 1992). The overt movement of the eyes during a saccade also appears to involve a momentary blindness, which is a form of interruption of visual activity. Of course, the immediate interruption of current processing is important to the survival of an individual who must respond appropriately when a new and possibly threatening object or event signals its presence.

Evidence from both visual and auditory investigations of the interruptive effects of strong orienting reactions may have several noteworthy implications. If the interruption of ongoing processing is a necessary effect of the sudden onset of an intense stimulus, then even if preparatory attention is already aligned to that particular stimulus, attention to its ongoing processing may be interrupted. Ordinarily, after a new stimulus initiates orienting, attention can be directed to the source of the new stimulus to evaluate or "realize" what and where the stimulus is. But if the first onset is quickly followed by another sudden onset, then the attentional processing of the first onset is interrupted. A sequence of strong sudden onsets would therefore seem to produce repeated interrupt signals, which compel the individual continually to reorient to the same source while making no progress in identifying or evaluating the stimulus. Some examples of stimuli that seem to produce this repeated state of "noticing without realizing" are modern sirens that fluctuate rapidly in pitch and loudness (in contrast to the more traditional sirens that produce a slow sweep between a high and low pitch) and the flurry of firecracker bursts at the end of a fireworks display. The repetition of intense stimuli appears to be an important ingredient of "hype."

The relation between intense sudden onsets of stimuli and sustained attention may be compared to that between a spark and the flame of a cigarette lighter. Once a spark ignites the flame, the flame develops quickly to its full potential. If a series of sparks is rapidly produced, however, then the flame initiated by the first spark will be continually interrupted and thereby prevented from developing to its full potential. Analogously, when someone shouts "Watch out!" repeatedly, it is difficult to realize what to watch out for. Instead of being *re*oriented, the listener may become *dis*oriented.

By extension, some degree of interruption of ongoing attentional processing may also take place in response to an endogenous cue at the moment when top-down processes activate a new object location. The duration of the interruption in the case of endogenous cuing and perhaps in the case of low-intensity exogenous cues may be very brief, so that its effects on measurements taken in the typical cued-location experiment are small.

In sum, it appears that visual orienting to a spatial location involves not only the eventual alignment of attention to a new source of information but the interruption of ongoing attentional processing during the early moments following the orienting stimulus. The extent and duration of the interruption depends upon the strength of the orienting signal at its onset.

## The "Peaked Distribution" of Attentional Activity

One way to conceptualize preparatory attention to a spatial location is to imagine a local distribution of activity in a spatial map (e.g., Downing and Pinker, 1985; Hendersen, 1991; LaBerge and Brown, 1989). In the LaBerge and Brown theory, the shape of the distribution is typically assumed to resemble a normal (Gaussian) distribution commonly referred to as the "bell curve," or back-to-back exponential distributions in the shape of a flat-topped mountain: the plateau portion of these shapes represents peak activity at the attended area, and the amount of activity decreases with distance from the attended area (see Figure 3.1, below). The size and level of activity of the attended area is presumed to be influenced mostly from top-down projections from working memory that temporarily store information from endogenous and exogenous cues, as well as from expectancies retrieved from long-term memory into working memory.

The height of the curve at the attended area is assumed to be affected weakly by recent presentations of objects at that location and strongly by momentary top-down attention to an object's location. The spread of preparatory activity outside the attended area is assumed to be determined by recent processing of objects that occurred in locations peripheral to the attended area. Thus, the extent of the momentary activity distribution is influenced by the short-term memory of activated locations. Evidence supporting some kind of short-term storage for locations comes from analyses of synapses in cortical and thalamic cells that contain metabotropic receptor channels (e.g., McCormick, 1992). These synapses are located on distal dendrites (dendrites away from the cell body), and a brief activation of the synapses decreases the firing threshold at other synapses near the cell body, where the lower threshold is maintained for at least minutes. The metabatropic receptors could therefore code the occurrences of recent events that have activated the distal dendrites of cells in a spatial map encoded in a cortical area and/or a thalamic circuit that serves that cortical area. Thus, the variation of threshold in cells that map the locations of recently presented objects could influence the range and the general shape of the distribution, while sustained attention to an object influences the height of the distribution's peak.

Because the notion of a short-term threshold shift is still being developed within the present theory, for the present it may make theorizing simpler if the lowering of threshold at an object's location is represented in the model as an increase in baseline activity.

The effects of recently presented stimuli on the shape and extent of the preparatory distribution were tested in a series of experiments that varied the spatial range of stimulus locations (LaBerge and Brown, 1986). The response-time V-curves (of the form shown by the upper curve of Figure 2.1) were produced over ranges that varied from approximately 2 degrees to approximately 6 degrees, while the width of the preparation cue at the center location was held constant at approximately one-third of a degree. The preparation cue could not have generated the activity distribution by itself because the slopes of the V-curve arms did not remain constant, but rather significantly and substantially decreased as the range of targets increased. Apparently the variable that changed the slope of the V-curve arms is the range over which stimuli of recent trials were presented (LaBerge and Brown, 1986, 1989), and therefore the spread of the activity distribution is assumed to increase with the range of recently presented stimuli.

The finding that the range of recently presented stimuli affects the time needed to shift attention from the center location to eccentric locations suggests that the locations of these recent stimuli have been stored in a short-term type of memory. The stored activity distribution is assumed to be latent until attention is activated by a cue, much in the same way that the activity level of items in long-term memory are latent until raised into working memory by an appropriate cue. In memory retrieval, items are generally retrieved more rapidly from the working memory state than from the long-term memory state. Similarly, it is assumed here that attention to spatial locations can be induced more rapidly when the cells coding for those locations have been elevated to a higher state of activity by an appropriate cue.

The activity distribution of preparatory attention is almost always wider than the current area of selective attention and endures for longer periods of time than the current area of selective attention. As trials of a typical spatial-location experiment accumulate, the activity distribution is assumed to approach a stable form. Hence a spatial cue presented on trial 50 will produce a different preparatory distribution than a spatial cue presented on trial 1 or 10. The height of the distribution at the attended area (typically the central peak) is presumed to be more volatile, since it rises quickly during the presentation of a central cue and decays rapidly after a target appears.

The distribution of preparatory attention does not itself selectively process a target object surrounded by distractors. Selective attention is presumed to be carried out by the momentary attended area. A new, momentary, attended area can be established at any location in the range of the existing distribution, and this is what happens when attention is aligned at a target following a cue. If the central location is cued and the target appears at the same place, then the preparatory activity here already shows high activation, and only a little additional activation is required to enhance the attended area above the activation of the immediate surround. On the other hand, if the central location is cued and the target appears away from center, then more additional activation is required to enhance the new attended area above the activation of the immediate surround. This line of reasoning provides the basis for predicting shifts of the attended area without analog movements through locations that are intermediate between the initial location and the new location. When attention shifts to a new location somewhere eccentric to the center of the distribution, the shift time is determined by the activity under the distribution curve defined by the area of the new location.

It can be shown that a distribution of activity with a peak located at the initial location of the attended area will generate a monotonically increasing curve of response time as a function of the distance that the attended area is shifted. One could invert the peaked activity distribution to illustrate this monotonic relationship. However, it can be shown mathematically that linear V-curves, which are most typically found in these kinds of experiments, could not arise from activity distributions constructed with back-to-back exponential functions, nor from activity distributions constructed with back-to-back linear functions. The mathematical function that can be shown (LaBerge and Brown, 1989) to generate linear arms of V-shaped response-time curves is:

$$RT(x) = C_i s(x) / g(x) + T_r$$

where $RT(x)$ denotes the response time for location of the second target, $g(x)$ denotes the value of the activity distribution at that location, $C_i$ denotes the degree of target/flanker similarity, $s(x)$ denotes visual sensitivity changes due to retinal eccentricity of the target, and $T_r$ denotes the residual time that may be attributed to factors such as conduction times and time needed to execute the response.

## Shifting Visual Attention through Space

*The moving-spotlight model of attention shifts.* Probably the most frequently used metaphor of attention is the spotlight or searchlight (e.g., Crick, 1984; Posner, Snyder, and Davidson, 1980; Norman, 1968; Treisman, 1980; Shulman et al., 1979). The spotlight can be viewed as the opposite of the filter; instead of inhibiting information flow in the surround of the attended area, as a filter does, a spotlight facilitates the information flow within the boundaries of the attended area. Its beam may be directed at locations in representations of sensory spaces (e.g., to somatosensory and auditory sites as well as to visual sites) and to locations in semantic fields (e.g., to categories, plans, motive sites, and so on).

The notion of a spotlight has wide reach. It implies that there is only one beam and that the size and intensity of the beam can be varied. In particular, when it is shifted from one location to another, it moves in an analog fashion, even if the beam is turned off while it moves (Briand and Klein, 1987). Also, when a beam shifts away from one location, it leaves that location in darkness.

The conventional mapping of the operations of a physical spotlight onto attentional operations of the cognitive system has been strengthened further by the fact that the fovea of the eye shares the major properties of a moving spotlight: it is a confined, compact area with specialized visual properties and it can be moved in a continuous manner around the visual field (by eye muscles or by the head or trunk muscles). Because the spotlight metaphor successfully captures the widely held belief that attentional operations are almost always concentrated within a compact processing area in the visual field (e.g., Heinze et al., 1994), it is tempting to accept the additional property that the critical area is moved intact in a quasi-continuous fashion across the visual field. When the beam is shifted away from a location, that location loses its attentional processing, and the new location begins its attentional processing only when the beam arrives. Thus, the spotlight metaphor combines in one structural mechanism the capability both to select a compact area of the visual field and to move the location of the selected area.

The spotlight metaphor is very seductive because some of its aspects apparently fit so well what we know about attention that we assume that all the other aspects also adequately describe attention. For example, while many researchers accept the properties of a bounded area of attention of a variable size, they question the ability of attention to move across a sensory or semantic field in an analog manner. Eriksen and Murphy (1987) estimated that, at the time of their review, the evidence favored no particular explanation of how attention shifts across the visual field. They commented that the accumulated evidence owed its ambiguity not so much to the heterogeneity of experimental designs but to our ignorance about the processes of attention. But since 1987 an increasing number of studies appear to oppose the view that attention moves in the analog fashion of a spotlight.

*The preparation/selection model of attention shifts.* A detailed account of how the present theory accounts for attention shifts in visual space requires as background an explanation of how a relatively broad distribution of preparatory attention, induced from processing a cue, interacts with a relatively narrow expression of selective attention, induced when a target is processed. The diagrams of Figure 3.1 show an activity distribution in the posterior parietal cortex (PPC), which corresponds to the levels of preparatory attention across the visual field. The peak portion of the distribution in the PPC (shown by the darker line) is produced by the gradual increase in attentional activity in the attended area at the center item of the warning signal. The subject is induced to *attend* (not merely to *expect* to see something) at the center location by presenting the upcoming target very briefly. Unless the subject prepares attentively at the center location, she will not accurately identify the target. The peaked distribution is sustained briefly by further attentional processing of the (first) target, S, in the display 585858S585858. The lower part of the activity distribution (shown by the thinner line) represents the memory of recent locations of targets, both first targets (located always at the center) and second targets (located across a horizontal range). This component of the distribution is activated from memory by the warning-signal cue. The upper part of the activity distribution represents the component that is added to the memory component when selective attention is concentrated at the center location during the warning signal.

The lower diagram of Figure 3.1 shows the state of the system after the second target has appeared and the identification of the letter R in the display VRV takes place (if the delay between the first and second targets had been long, the preparatory attention peak would have decayed toward the level of the memory component of the activity distribution). The VRV display is presented to the left of center, near the edge of the range of recently presented targets. When the

**During Display of Warning Signal**

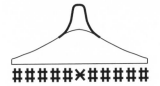

**During Display of Target #1**

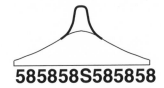

**During Display of Target #2**

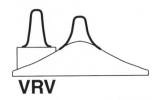

*Figure 3.1.*   Theoretical representation of sustained preparatory attention and transient selective attention in a spatial map within the cognitive system during a shift of attention from a target, S, at the center location to a target, R, at a location to the left of center. Locations of targets are shown in a horizontal array, and the corresponding amount of neural activity is indicated by the height of the graph lines. During the warning signal preparatory attention is built up at the center location, ✱, in anticipation of the brief display of the letter S (Target #1), and the resultant activity is sustained into the time that the letter R (Target #2) is processed. The time to build selective attention to the letter R (represented at the left side of the bottom graph) depends upon the amount of preparatory activity already existing at its location. Preparatory attention to the middle item in the upper diagram is induced over time from prefrontal processes, and selective attention to R in the three-item-wide stimulus, VRV, is induced briefly by prefrontal processes. The lighter, broadly distributed graph lines represent the memory of recently presented locations of targets.

VRV appears, visual information is registered in the occipital projection area (area V1) and is projected to the PPC, where it raises the activation corresponding to the width of the three-letter group, VRV. The information from the three-letter group is also projected onto the dorsolateral prefrontal cortex (DLPFC), which maps spatial locations, and there it interacts with remembered instructions (in working memory) to identify the center object. The resulting activity at the center location within the DLPFC map is then projected back to the activity distribution in the PPC, where it raises the existing activity of the VRV object at the center location and produces the small peak located in the left part of the distribution. The projection of activity at this location continues until it accumulates a higher activation value than any other location represented by the activity distribution. The elevated activity at this new location represents the attended area after attention shifts from the first target to the second target.

The goal of concentrating attention at the center location of the VRV display is the identification of the letter R, but the additional structures needed for concentrating attention in feature maps of the brain will be described in later chapters. As a brief preview of that later account, it can be said here that the projection from the PPC to the pulvinar nucleus of the thalamus will induce the pulvinar to output a relatively narrow flow of information onto the pathway of information flowing from V1 to circuits that identify the letter R in the inferotemporal cortex (IT). The result of the pulvinar output's interaction with the V1-to-IT pathway is a modulation of flow that is regarded as the expression of attention to just the center location of the stimulus, VRV. The information from the center location subsequently enters the IT circuits and the letter R is identified.

If the VRV stimulus had appeared nearer to the center location or at the center location, instead of at the extreme left, then the rectangular column rising out of the activity distribution of the PPC in the lower diagram in Figure 3.1 would be shifted toward the center. Because the preparatory-activity distribution is higher nearer the center location, however, less additional activation from the DLPFC is needed to raise the activation at the location of the R object to a suitable dominant value. In this manner, the differences in response time of attention shifts, indicated by the V-curve, are determined by the time taken by the DLPFC to project the required amount of activity to the PPC distribution.

Thus, the present account of shifting attention in space is based on two manifestations of attention. The preparatory manifestation is expressed as a relatively slowly decaying distribution of activity in a location map of the brain that is superimposed on a wider and more enduring distribution of activity that expresses the short-term memory of recent locations. The selective manifestation is expressed as a relatively narrow and very brief distribution of activity that is shaped more closely to the size of the target stimulus. Under usual circumstances a brief selective operation will have little effect on the shape of the preparatory activity distribution. An example of longer-lasting preparatory attention to one location while selective attention is briefly concentrated elsewhere is the cognitive processing of a basketball player, who sustains strong preparatory attention to the location of the basket while repeatedly employing selective attention elsewhere in brief bursts as he maneuvers the ball between opponent players.

*Experimental indicators of attention shifts in visual space.* The time taken to respond to a target in a location away from the present location of attention has been shown to depend on the distance that attention is shifted (e.g., Downing, 1988), and the relationship between distance and time appears to be approximately linear (e.g., Tsal, 1983; LaBerge and Brown, 1989). The linear relationship suggests that the time required to shift attention from one location to another could be the simple result of an analog trajectory mechanism, in which time to move to a new location depends upon the distance moved. Considerable research has been generated by the hypothesis that attention shifts are the result of a mechanism that produces continuous movement across an internal map of external space; this section describes experiments that test this hypothesis in the context of an opposing hypothesis that attention shifts in a discrete manner that is independent of the distance involved.

To measure attention shifts, the experimenter typically induces the subject to attend first to a particular location (e.g., at a fixation cross in the center of the visual field, or at some item on a clock-like circle around that center). Then a stimulus is presented that induces the subject's attention to shift to a second location at a particular distance from the cue. For example, in the two-to-go experiment used by LaBerge and Brown (1989), the subject was instructed to attend rather intensely to a central character in a horizontal string

of characters (e.g., the letter S in 585858S585858). The high similarity and close proximity of the target character and the distractors
were intended not only to control the location of the attended area
of preparation at the center but also to control the size of the attended area to one character of the string (see Table 2.1). On half
of the trials, the center item was replaced by a similar item (e.g.,
5858588585858 or 5858585585858). Immediately following this
display (the first target), a second target appeared somewhere along
the same horizontal line on the monitor screen. Typically there were
five alternative locations in which the second target appeared, one
at the center and two on each side of center, making a row of five
equally spaced location points.

Each of these five locations represents a destination point of the
attention shift from the center location (with the center location
of the second target serving as the control condition of zero shift
distance). To ensure that attention is indeed shifted to a particular
destination location, a target must be selected from closely positioned distractor items (e.g., the letter R in VRY). Half of the time
the letters P and K are substituted for the letter R, so that the subject
must attend closely to the letter target to identify it correctly by a
conjunction of two critical line features (e.g., Treisman and Gelade,
1980). The task for the subject was to respond with a button press
only when an S (always located at the center) was followed by an R
(in any of five locations).

The results of this type of experiment are shown by the V-shaped
curve of response time in Figure 2.1, where the data are provided
by six subjects. The other curve in this figure, which has almost a
horizontal shape, depicts the response-time results of a variation in
this procedure that induces preparatory attention to a spatial width
of a five-letter word instead of to the width of a single letter (LaBerge
and Brown, 1989; LaBerge, 1983). The V-curve and horizontal curve
demonstrate a range of slopes that are presumably determined not
only by the location of the attended area before it shifts, but also by
the size of the attended area before it shifts. The width of the central
peak of the preparatory distribution (see Figure 3.1) is induced by
the subject's expectation of the size of the first target (a letter or a
word). If one were to omit the first target from the sequence of trial
events, then the size of the attended area would be free to vary at
the whim of the subject, and the resulting response-time curves

would show more variability and the slope values would lie somewhere between those shown by the two curves in Figure 2.1.

The approximately linear arms of the V-curve in Figure 2.1 can readily be accounted for by a moving-spotlight model of attention shifts: the attentional spotlight is initially positioned at the center location because the first target is processed there, and then the spotlight moves at a constant velocity to the location of the second target when the second target appears at a location away from center. The horizontal curve in Figure 2.1 can be accounted for by assuming that the expectation of the size of the first target increases the width of the spotlight to accommodate a five-letter word, and when the second target appears, the spotlight narrows to the width of a single letter. The spotlight model can easily add the assumption that the time to reduce the size of attention is constant, so long as the smaller size is not outside the initial beam.

The briefly flashed target is more likely to be accurately identified when adequate time is provided to build up the peak of preparatory attention at the attended area during the warning signal. In the two-to-go experimental procedure shown in Table 2.1, the warning signal serves as a cue, because it marks the center location and size of the upcoming target letter, and accurate identification of the first target that follows the warning signal guarantees that the subject aligns preparatory attention to the location and size of the center character of the warning signal. Thus the warning signal serves as the cue that guides the buildup of the activity distribution's peak.

During the 1,000 msec warning-signal display, the subject has the opportunity to achieve a high degree of attentional preparation for the location and size of the first target (the letter S in 585858S585858). The inducement not only to expect but also to prepare attentively for the location and size of the letter S is the very brief duration of the 585858S585858 display. Without precise prealignment of the attended area, the S will often be missed, because the subject does not have enough time once the display appears both to adjust to its location and to identify the center letter. Thus, when the S has been identified, one can be reasonably assured that attention is narrowly constrained to the size and location of the center letter. Then, when the second target appears in one of the five locations and attention shifts to its location, one will have a clear notion of where the shift begins.

*Experiments that vary target/distractor similarity.* A similar experiment was designed specifically to test the moving-spotlight model against a preparation/selection model. This experiment employed the same procedure as the one just described but varied the similarity between the second target and its distracting stimuli (LaBerge and Brown, 1989). The results of this experiment, based on 12 subjects, are shown in Figure 3.2. The observed significant differences in slopes produced by changes in target/distractor similarity can be accounted for by a moving-spotlight model only if the additional assumption is made that the velocity of the spotlight movement was changed according to the difficulty induced by the similarity of each type of target/distractor. This type of adjustment seems remote, but

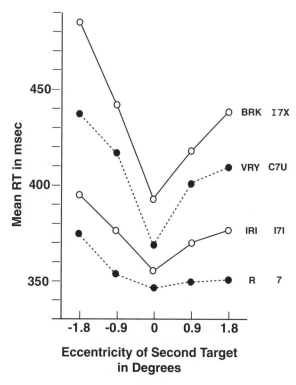

*Figure 3.2.* The effects of target/distractor similarity on time required to respond to a target (R or 7) after the subject had already identified a first target (the letter S) at the center of the visual field. In each block of trials, the target was flanked by the same distractors.

it is possible, given that the data were collected in a blocked design, in which each type of distractor was varied across days. Thus, a subject could use foreknowledge of the target/distractor difficulty to adjust the velocity of spotlight movement before each block began and maintain that velocity throughout all the trials of that block.

In a subsequent experiment, foreknowledge of the target/distractor difficulty was prevented by varying distractor types *within* a block of trials, so that the subject had no way of predicting which type of distractor would appear on a given trial and therefore could not adjust the velocity of attention shift accordingly. The results of this procedure, based on 20 subjects, are shown in Figure 3.3. According to the moving-spotlight model, subjects can either set a constant velocity for all trials, or vary the velocity in a manner that could not be correlated with the upcoming target. In either case, the moving-spotlight model predicts that the data should show equal slopes across all distractor conditions. The data showed slope differences

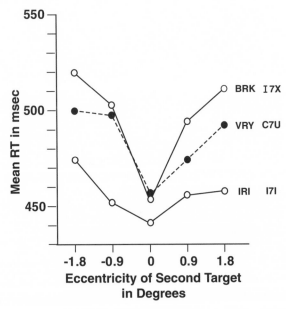

*Figure 3.3.* The effects of target/flanker similarity on time to respond to a target (R or 7) after the subject had already identified the letter S at the center of the visual field. In each block of trials, the target was flanked by random distractors.

that were again significant and systematically correlated with difficulty of the distractor, however, and these differences cast serious doubt on the ability of the moving-spotlight model to account for the shift of attention in these experiments.

The results of the foregoing experiment suggest that the target/distractor similarity variable interacts with some process other than the movement property of a spotlight. A survey of the properties of the spotlight metaphor itself (see Chapter 2) does not suggest a process that would account for these results. It is not adequate simply to assume that attention is spread over the entire range of targets, perhaps with more concentration of attention in the center, somewhat like the distribution of clarity produced by a zoom lens. (In the preparation/selection model, this kind of attentional property is contained in the distribution of preparatory attention but it accounts for only part of the attentional processing in these experiments.) There must be an additional specification of how a particular target undergoes selective processing to separate its information from the distractor information, particularly when the distractors are close to and similar to the target.

Perhaps the moving-spotlight theory could be saved by assuming that there is really no shift of the location of the attended area at all in these experiments, but only a very rapid expansion and contraction of the size of the attended field. After the first target letter is processed with a narrow focus, the size of the attended area rapidly zooms outward, and when the second target appears it zooms inward again to the width of the target letter. No movement of the attended area took place, because the narrowing of the attended area to the second target occurred within the broader attended area that had been quickly generated between the first and second targets.

The problem with this account involves the observed changes in response-time slopes with varying target/distractor similarity within a block (see Figure 3.3). On any given trial, the subject has no way of knowing how to adjust the shape of the broadly focused attention established between the first and second target in a way that would produce the particular slope corresponding to a particular target/distractor similarity of the second target on that trial. Therefore, the moving-spotlight model modified by the assumption of a zoom lens apparently cannot account for the present data.

*Experiments that do not begin the attention shift from a location having high preparatory attention.* In an attempt to provide a simpler and more direct test of the moving-spotlight model of attention against the preparation/selection model, the two-to-go procedure that was described earlier was modified to present three successive targets within a trial following a warning signal (LaBerge, Bunney, Carlson, and Williams, submitted). The extra target was inserted between the first and second targets, so that the targets are now numbered in order of their appearance: the first target appears always at the center, the second target appears either at the center or at the extreme left or right locations, and the third target appears in one of five locations (spread over a range of approximately 1.7 degrees of visual angle). The trial events of this three-to-go procedure are shown in Table 3.1. The second target, the triplet VRV, is presented equally often either at the extreme left location, at the center location, or at the extreme right location.

The purpose of the second target was to provide three locations from which the subject's attention could begin its shift to one of the five locations of the final target. Although the center location is assumed to have a high degree of attentional preparation, induced by the expectation of a briefly presented first target, the two outside locations are assumed to have a low degree of attentional preparation. The close spacing of the distractors to the second target should insure that even a zoom-lens version of the moving-spotlight model would assume that the zoom lens had to be concentrated around the center letter R in order to eliminate the information arising from the V distractors (e.g., Eriksen and Webb, 1989).

*Table 3.1*

| | |
|---|---|
| Warning signal | # # # # # ✱ # # # # # |
| Target #1 | G Q G Q G O Q G Q G Q |
| Target #2 | V R V |
| Target #3 | G O Q |
| Locations of target #2 | ● ● ● |
| Locations of target #3 | ● ● ● ● ● |
| Catch trials for O were C, Ø | |
| Catch trials for R were K, P | |
| Instructions: Press the button only when the sequence O–R–O occurs | |

Subjects were instructed to respond with a button press only if the first target was an O, the second target an R, and the third target an O. In the experiments employing the two-to-go procedure that were described earlier in this section, the attended area presumably always began its shift to the final target from the center location. In the present experiment the shift to the final target could begin from the right extreme or left extreme location as well as from the center.

*Predictions by the two models.* The response-time curve predicted by the moving-spotlight model has a V-shape when the second target is located at the center location. When the second target is located at either the left or right end of the horizontal range of targets, however, the moving-spotlight model predicts a linear curve that has its minimum at the location of the second target and its maximum at the opposite end of the five-location range. If one were to plot the curves for the cases in which the second target were at the left and at the right ends of the range, the resulting plot would take the form of an *X.*

The response-time curve predicted by the preparation/selection model has a V-shape regardless of the location of the second target, because the time needed to shift attention for this model depends on the activity distribution, not on the momentary location of the attended area (see Figure 3.1). One would expect that the slope of the V-curve arms would be greater for the case in which the second target is at the center location, because the processing of the target in that location sustains the high activity peak at the center for a time. When the second target is located at one of the extreme ends of the range, the activity at the peak is expected to decay while the second target is displayed. At the time that the third target appears, therefore, the activity distribution will be less peaked, and the resulting arms of the V-curve should show a lower slope than when the second target is located in the center.

The results of this experiment, based on 15 subjects, are shown in Figure 3.4. Each of the three curves represents the response times given the location of the second target (T2), which immediately preceded the third target. All three curves showed V-shapes. In particular, for the two curves showing results for the extreme left and right locations of the second target, the slope of the arm nearest the location of the second target was tested against a zero value statistically, and the difference was found to be highly significant.

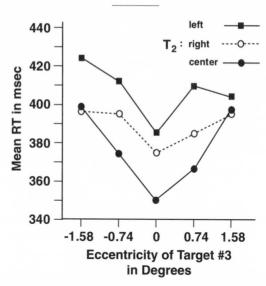

*Figure 3.4.* Mean response time involved in identifying a third target after the subject had already identified a first target located at the center of the visual field and a second target (T₂) located either at the left, right, or center of the field. The preparatory state for the first target was high.

The straightforward interpretation of these results favors the preparation/selection model over the moving-spotlight model, because the major determinant of the time required for an attentional shift appears not to be the momentary location of the attended area but rather the level of activity at the location to which the shift is made. This level of activity at various locations, represented by the preparatory activity distribution (see Figure 3.1), persists while selective attention moves to other locations.

*A review of the preparation/selection model.* The three experiments that have just been described apparently support the hypothesis that the shifting of attention in visual space involves two main processes: a broad distribution of activity that endures over relatively long time periods induced by preparatory attention, and a relatively narrow distribution of activity induced by selective attention that decays rapidly. The first process involves the combination of two component distributions of activity: a broad distribution that is based on the memory of recently presented targets and a more concentrated distribution that is based on the buildup of attentional concentration during a cue, which is regarded as a manifestation of preparatory attention. The second main process involved in shifting attention is the selection of a target at a new location, which is assumed to be accomplished by a short-lasting, narrow distribution of attention that is projected to pathways of featural information, where it modulates information flow so that target items can be identified when distractors are nearby. Thus, the expression of attention both as preparation and as selection is assumed to be relatively narrow distributions of activity. When the narrow activity distribution is manifest as preparation, it typically builds up over time to a high concentration (and combines with a broader spread of preparation based on a lower threshold of activation), and when the voluntary control of the attention process is subsequently withdrawn (as when attentional activity is initiated elsewhere) the preparatory-distribution peak decays rapidly and leaves a residue that endures over time as a memory of that location. The residue then becomes part of the stored threshold distribution. On the other hand, when the narrow distribution is expressed as selection, the buildup of attentional concentration is typically just high enough to produce an attended area of the size of the target object, and concentration remains at this level only long enough to enable the identification of the target in the V1-to-IT pathway.

The two attentional distributions corresponding to preparation and selection account for attention shifts by assuming that the two distributions can exist at two different locations at the same time. The selective distribution modulating information flow at a spatial location need not correspond to the location of the peak of the activity distribution generated by preparatory attention. The selective process operates on featural information entering an identification module while the preparatory process (along with the stored activity distribution) determines how rapidly the selective distribution can increase to an effective level at a new location. In contrast, the moving-spotlight model has difficulty separating these two functions because it assumes that there is only one "beam" operating at only one location at a time: if the beam is to operate at a new location, it must depart from the old location, without leaving a trace or residue as a record that it has been at the old location.

A critical assumption for the preparation/selection account of these data is that the decay of the selection distribution is much faster than the decay of the preparation distribution. A more direct test of this assumption is desirable, for which we turn to the two variables that likely influence the decay rate of attentional activity: its duration and the level to which it decays.

Preparatory attention to a center location, represented by the thick-lined curve in the upper diagram of Figure 3.1, is built up and sustained over the 1,000 msec duration of the warning signal in anticipation of the upcoming brief display of the first target. This duration exceeds by many magnitudes the duration of the selective attention to the second target, shown in the lower diagram of Figure 3.1 (and to the first target, shown in the middle diagram). Longer durations of preparatory attention may produce stronger modifications of synaptic structures that require longer time to reset to normal.

The baseline activity on which preparatory attention is assumed to be built is illustrated in Figure 3.1 by the thin-lined curve. This activity, which represents the stored effects of many recent trials, is higher at the center location than at the other locations. When preparatory attention on a given trial is generated at the center (as in Figure 3.1), it does not decay as much as it does when preparatory attention is generated at a location away from the center. As the activity level approaches its baseline, it may decay more slowly.

In summary, two types of experiments were designed to test the assumption of the moving-spotlight model that attention shifts across the field in a continuous or analog manner, so that the distance of the shift affects the time required to complete the shift. A preparation/selection model, on the other hand, bases the time needed to shift attention on the amount of preparatory activity at the destination of the shift. Because the amount of preparatory activity at a location has no necessary relationship with distance of a shift, the time required to shift attention is predicted to be independent of distance. In many laboratory situations, however, the highest level of preparatory attention exists at the currently attended location, particularly when the currently attended location is at the center of the visual field. Hence when a cue or a target stimulus induces a shift away from center, the amount of preparatory attention at the new location is almost always less than the preparatory attention at the center location. Since the gradient of activity slopes downward from the center location to the new location, the predicted shift time increases with distance from the center location in much the same manner as predicted from a moving-spotlight model.

This state of affairs makes it difficult to test the moving-spotlight model with simple experimental designs. Hence the two types of experiments described in this section, which involve relatively special conditions (e.g., a high level of attentional preparation, variations in target/distractor similarity, etc.), were designed to compare contrasting predictions by the moving-spotlight model and the opposing preparation/selection model.

*Attempts to measure the decay of the preparatory activity distribution over time.* The overall shape of the activity distribution is determined not only by the locations of recent targets but also by the momentary effect of the cue, which in the present experimental examples is located at the center of the range of stimuli. It is assumed that the amount of attention concentrated at the cued location determines the amount of activity at that location. To some extent the activity at the cued location spreads to neighboring locations, but the spread is much narrower than the spread produced by the range of recent target stimuli. Furthermore, the peak of the distribution at the cued location will decay unless sustained by selective attention to objects in the same location. Thus, the momentary shape of the activity distribution at the time that the subject attends to the cue is determined by the level of attentional concentration to the cue and the memory of recent target locations.

When attention is withdrawn from the location of the cue, the central peak of the activity distribution is assumed to decay, while the tails of the distribution remain relatively unchanged. In a two-to-go experimental procedure, the warning signal acts as the cue that guides the alignment of attention to the center item. During the 1,000 msec that the warning signal is displayed, attention to the center item of the warning signal builds to a level of activity that is under the voluntary control of the subject. Then the warning signal goes off and the first target appears in the center location where attention has been concentrated. Identifying an S with similar distractors nearby (5 and 8) requires concentrated selective attention, and either this attentional activity raises the peak of the activity distribution to a higher level or, if the peak was already at a high level by the end of the warning signal, the attention given to the first target, S, sustains the peak at its level for a brief additional time period. After the first target goes off, the second target appears and the momentary activity distribution determines the time required to build a brief narrow peak at the location of the second target.

An experiment with four trained subjects varied the time between the onsets of the first and second targets in an attempt to measure the decay of the activity distribution. The first target was a string of letters, 58585S85858, and half of the time the S was replaced with 5 or 8; the second target was VRV, and half the time the R was replaced with a P or K. The instructions were to press the button only if S (which appeared always at the center location) was followed by an R (which could appear at any of five horizontal locations). Since the interval between first and second target onsets (the stimulus onset asynchrony, SOA) is made up of the duration of the first target plus the interstimulus interval (ISI), we varied both components of the interval orthogonally. The four durations of the first target were 67, 100, 200, and 500 msec, and the four durations of the ISI were 50, 150, 300, and 500 msec. The duration of the warning signal was 1,000 msec, and the duration of the second target was 200 msec. The obtained slope values averaged across the two arms of the response-time V-curves are shown in Figure 3.5.

The slope data show a decrease with time between the onsets of the first and second targets that is rapid at first and approaches an asymptote, which is the type of trend expected from a decay process. The absolute response-time data from the second targets are complicated by the fact that the processing of the first target extends into the time that the second target appears, so that the processing of the second target may not actually begin at the onset of the stimulus. The slope measurements are assumed to be relatively independent of the various factors that affect absolute response time, and they suggest that, following preparatory attentional concentration at the center location, the activity value of the central peak of the activity distribution decreases relative to the values at the tails of the distribution.

Thus, in accounting for attention shifts, the present model employs both the preparatory and selective manifestations of attention. Attention to a cue activates the memory of recent target locations, represented by a relatively broad spread of activity, and also produces a momentary peak of activity at its own location, representing preparatory attention. When a target appears at a new location, se-

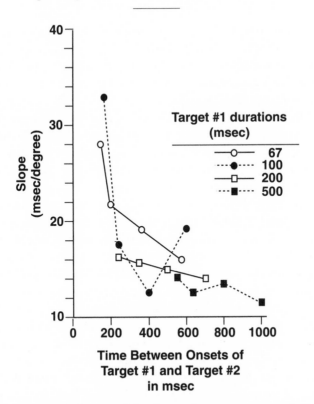

*Figure 3.5.*   Putative indicators of the decay of preparatory attention: average slopes of the response-time V-curves determined at varying intervals following the onset of the prepared-for stimulus (the letter S) located at the center of the visual field.

lective attention separates target information from distractor information, but the rate at which this selection occurs is influenced by the preparatory activity already existing at the new location. When preparatory attention is greatest at the center, the further from center that the cue or new target appears, the less the preparatory activity existing at that location and the longer it takes to induce selective attention to a target at that location.

## The Resource View of Preparatory Attention

It has been said that when one "attends to" an object one is "allocating resources to" that object. That is, the notion of attending to a particular stimulus property or to a response in anticipation of their imminent occurrence readily invokes the metaphor of allocating some kind of mental resource in this attending. Some aspects of the preparation process have been viewed by researchers as strongly resembling resource allocation. For example, while one attends to a spatial location at which a stimulus may appear, the preparatory effects seem to spread into the neighborhood of the anticipated location (e.g., Downing, 1988), suggesting a flexible allocation of attentional resources within the spatial domain. In some tasks it may even seem that attentional resources can be shared across two or more objects, and this hypothetical state of affairs can be easily represented by a kind of hydraulic model of resource sharing, described below. In particular, it may seem that the activity distribution described in this book resembles a distribution of resources over a neural map within the brain. This resemblance will be discussed after a description of how the notion of resources is applied by current investigators to attention experiments.

Attention theories borrowed the notion of capacity from physics, and from Shannon's information theory in particular (Shannon and Weaver, 1949). Transmission of information from stimulus input to response output occurs along a "channel," and physical channels have limited capacities in terms of the rate of flow; for example, in any hydraulic system water can flow only so quickly through a channel of a particular diameter. By analogy, information may travel only so quickly through neural pathways. Early forms of resource theories assumed a single channel with a non-specific reservoir (e.g., Welford, 1967; Craik, 1947), and resources could be allocated in various amounts as needed by the tasks at hand. Kahneman (1973) described two types of attention models involving resource limitations: (1) structural limitation models, which predict that concurrent tasks interfere with each other because the same mechanism is required to carry out two incompatible operations, and (2) capacity limitation models, which predict that concurrent tasks interfere with each other because the requirements of the two tasks exceed available resources. Later theories spelled out in more detail how multiple pools or reservoirs might serve different system structures (e.g., Navon and Gopher, 1980; Wickens, 1984).

The limitation of resources was a notion familiar to economists as well as to information theorists, and theories of optimization in economic production seemed suggestive of the strategy of the ideal observer in signal-detection theory, which could be used as a normative basis for analyzing the attention process. Optimization equations in search theory for World War II submarines developed by the mathematician B. O. Koopman (1957) served as a basis of the quantitative theory of optimal allocation of cognitive resources developed by M. Shaw (1978). The notion of optimization was also explicitly adopted by Sperling (1984) to determine how resources are shared across concurrent tasks.

The concurrent task has served as a representative experiment for many theories that assume attention is allocated as a continuous quantity across tasks. Subjects are instructed to divide attention according to some instructed proportion. In a recent study of resource allocation across concurrent tasks (Bonnel et al., 1987; Possamai and Bonnel, 1991), subjects were required to match the lengths of two pairs of lines presented to the right and left of a central fixation point. They were instructed to allocate attention to the two pairs of lines in the proportions of 0.80/0.20, 0.50/0.50, and 1.00/0, and the resulting $d'$ (a measure of the strength of a signal above background noise within the observer's system) estimates increased with the allocated proportion of attention, while ß (the response criterion) did not change. These results were nicely fit by Luce's model (Luce, 1977; Green and Luce, 1974), which makes the assumption that the attentional process samples one pair of lines with a frequency determined by probability, $p$, multiplied by the number of observations, $N$, and the other pair of lines with a frequency $(1 - p) \times N$. This assumption implies that attention is shared across the locations of the two pairs of lines. The possibility remains, however, that attention was not shared across the two locations but was dedicated to only one at a time and shifted rapidly between the locations. Attempts to distinguish these alternative hypotheses continue (Miller and Bonnel, 1994).

It seems that most concurrent-task studies, as a class, produce data that can be interpreted not only as a simultaneous sharing of resources or as an averaging over trials in which attention is allocated in an all-or-none manner, but as a probabilistic distribution between the tasks. Furthermore, even if data support sharing of resources within a trial, attention can shift rapidly during and following a stimulus display, which means that both the number and duration of attentional samplings can be distributed in such as way as to give the appearance of sharing.

Thus, opponents of the resource view of concurrent-task performance would maintain that the term *divided attention* is an instruction given to subjects, not a description of an actual sharing or spreading of attention, while the term *shifting focused attention* appears to serve appropriately both as an instruction and a description of the way people actually process information. Again, the blurring of the distinction between the anticipation or preparatory aspects of a task and the selective aspects of a task seems to have led to a confusion in the way resource allocation is applied to these tasks as an analytic device.

As a metaphor, resource allocation appears to have implications that do not square with the way in which selective attention appears to operate, both spatially and temporally. Instead of activation being distributed or spread among alternative stimulus items, in selection the spread of activation is constrained or blocked in order to sharpen the boundary separating the target information from the information in the surround. Allocation of resources would seem to take time, and in the brief moment that selection operates in some tasks (such as the identification of an object in a field of distractors), little time is available for carrying out the cognitive processing involved in allocating resources to stimulus locations or features.

When applied to the preparation manifestation of attention, however, the resource metaphor may yield more appropriate implications. Attentional preparation of the processing of locations or attributes of an object would seem to be capable of being modulated in intensity, and the preparation effect apparently spreads to locations and attributes that are similar to the particular anticipated object. Furthermore, in most experiments, ample time is provided prior to stimulus onset for cognitive (top-down) processes to assist allocations that may occur during preparatory attention. It is during the more extended periods of time prior to the onset of the stimulus that "feelings of effort" associated with attention (Kahneman, 1973) are experienced.

How is the resource view of attention related to the present view of preparatory attention as a distribution of activity in a brain "map"? Apparently the assumption of a limited capacity in a communication channel, which was a historical root of the resource notion, is not supported by evidence for a limited capacity in neural systems that process information about the location and features of objects. In fact, the notion that resources are limited was given up by many psychologists some time ago, so this particular issue is apparently not an important one in comparing brain activity distributions and resource theory.

The most serious problem with the resource view seems to lie in the lack of clear computational implications, especially when it is employed as an explanatory device. Allocating more or less fuel for an automobile engine does not indicate how the fuel is used to turn a crankshaft at varying velocities. When data from an attention experiment are explained simply by assuming that resources are allocated broadly or narrowly to some part of the visual field, exactly how a given amount of resource allocation is converted into the computations that are carried out by attentional processing, such as filtering out distractors or identifying objects and their attributes in disjoint areas of the visual field, is often not specified. On the other hand, when attentional activity is described in neural terms there are clues from neuroanatomy and neurophysiology as to how this activity can and cannot affect the computations assumed to be involved in what attention accomplishes when it aids the filtering of distractors and promotes the rapid identification of items and their attributes in separate areas of the visual field. The attention literature contains several examples of models for behavioral attention tasks that have been constructed in the context of neurobiological knowledge and then tested in psychological laboratories (e.g., Posner, 1980; Treisman and Sato, 1990). An example of a neural-inspired model of attentional concentration that has undergone tests using behavioral response-time measures will be described in Chapter 6.

## Maintenance Attention

In consideration of the apparent lack of experimental research directed to the maintenance manifestation of attention, the brief treatment of maintenance attention here will be more impressionistic and theoretical than the discussions of the selective and preparatory manifestations. It is hoped that this sketch of a new area of investigation will begin the work of integrating maintenance attention into the larger conceptual framework of this book and promote the design of experiments that can illuminate in detail its operations and controlling influences.

During the successive moments of our daily life, attention is involved not only in selecting an object out of a cluttered field for the purpose of judging it in some way (selective attention) or in anticipating the appearance of an object for the purpose of responding effectively to it (preparatory attention), but also in simply observing an object with no apparent immediate goal in mind. In contrast to the behavioral goals served by selective and preparatory attention, maintenance attention appears mainly to serve experiential goals. The example of consuming a reward has already been described as sustaining attention to ongoing sensory inputs without expectation of an upcoming event. Other situations in which attention often appears to follow a changing stimulus without a concurrent anticipation state are observing ocean waves, flames leaping in a fireplace, a bird in flight, a series of pictures in an art museum, or pages in a magazine, and listening to music or a lecture. In each of these examples, of course, one may convert the maintenance aspect of attention into a preparatory aspect by evoking a representation of an upcoming event, so that one is attending less to "what is" and more to "what will be."

When we attempt to formulate the computational goal of maintenance, we are led to conclusions that may be less satisfying than those drawn for computational goals of attentional selection and attentional preparation because the goal of maintenance appears to be attending to a process simply for its own sake. One may conjecture that giving attention to the taste of food or to the feelings of relaxation during a massage contribute to the well-being of the person, and that in consequence the person is better able to cope with the next challenges that confront him or her. The appeal of savoring certain experiences by giving them our full attention is undeniable, yet we apparently find it difficult to frame a justification for this use of attention in computational terms. To the question what computations are being performed during maintenance attention, the answer given here is that information arising from the attended object is enhanced relative to information arising from other sources, including other sensory modalities. But the immediate or future benefit of such an enhancement of information is neither specified nor implied in the expression of attentional maintenance, as it is for attentional selection and attentional preparation. Thus, if we were to program attentional operations into a robot, we would find good reason to include them for the purposes of selecting and preparation but, at the present state of our knowledge, we would see no clear purpose in instructing a robot simply to maintain attention on an object.

The expression of the maintenance aspect of attention is assumed to be enhanced activity in the cortical areas serving the object of attention, as is the case for the selective and preparatory aspects of attention. The maintenance activity is established and sustained from two sources: internal cognitive processes and external stimulus inputs. External stimuli typically initiate maintenance attention when they change abruptly in some way, as when a sudden movement or sudden change in luminance or color draws one's attention. As noted for preparatory and selective aspects of attention, however, the effects on attentional enhancements in the posterior cortex arising from changes in external stimuli are almost immediately overlaid by stronger influences produced from internal cognitive areas. Thus, the prevailing control of maintenance attention appears to arise from cognitive areas.

Although maintenance and preparatory aspects of attention seem similar because of their typically long durations (relative to the typically short durations of selective attention), there is a major difference between them: the presence of an associated expectation in the case of preparatory but not in maintenance attention. When in a state of anticipation, one sustains attention because an upcoming event bears interest. When in a state of maintenance, one sustains attention because the interest is being served by the sustaining of attention without regard to an upcoming event. Thus, sustaining attention on a cue word in an attempt to retrieve an item from memory would seem to be attention of the preparatory type, because the person anticipates the event of retrieving the target item while he or she is attending to the cue word. Quite possibly one could even conceive of the cue as pre-activating part of memory processing in somewhat the same way as one conceives of a stimulus cue as pre-activating part of the processing of the upcoming stimulus in sensory preparations.

The sustaining of preparatory attention is governed by a representation of the associated expected event or object in working memory; the sheer maintenance of attention requires no comparable governing factor in working memory. If maintenance attention is to be sustained over time, therefore, either the external stimulus must continue to change or the interest in the object or event must be sustained. A bird in flight can hold our attention for a time, but a bird perched unmoving cannot, unless the bird is intrinsically interesting for its visual features (shape or coloring) or for its endearment as a favored pet.

While interest is sustained, so will attention be maintained, as James pointed out (James, 1890). When interest begins to wane, attention will be particularly susceptible to capture from other sources. Thus, the major controlling source of maintenance attention would seem to lie in the anterior cortical areas where motivational processes influence attentional operations more directly than in posterior cortex (this point will be supported by neuroanatomical evidence later in the book).

The external control of maintenance attention is customarily viewed as involuntary while the internal control is viewed as voluntary. William James regarded the internal control on this kind of attending as a kind of mental action that was closely linked with the "will" (James, 1890). It is fitting to note here that James's philosophical crisis at age twenty-eight found its resolution when he realized that he had the *choice* of continuing to attend to something or not (Myers, 1986; Perry, 1954). He had just run across a statement by Charles Renouvier that free will is "the sustaining of a thought because I choose to when I might have other thoughts." James believed that this recurring opportunity to choose to continue a thought was not an illusion but rather a genuine opportunity to exhibit "free will," and this belief was apparently instrumental in easing his way through a serious personal crisis. This belief in individual choice framed at the primitive cognitive level of momentary attention found its way into higher-level cognitive philosophical beliefs: it is an element of the philosophy of pluralism, in which the unfolding events of the world are new and unpredictable, as opposed to monism, in which the unfolding events of the world are derived from an unchanging, universal scheme of some kind.

The present descriptive account of maintenance attention is consistent with James's claim that the attention process makes it possible for the thought of any object to prevail stably in the mind. More than once he suggested that voluntary attention refreshes a thought that, if left to itself, would decay.

Synonyms for maintaining attention are *regarding* and *observing*, when the object of attention is an external stimulus, and *considering* and *entertaining*, when the object of attention is an internal cognitive item or event. These terms suggest the sustained character of maintenance attention. Another possible synonym of attending, *detecting*, implies a brief rather than a sustained process, and the synonym *noticing* was used earlier to denote a process that takes place following an orienting cue, in which top-down control of sustained attention may be suspended. An example of maintenance of the observing kind is the annoying attention one gives to a noise in the environment, such as the honking of a horn outside a library window or the ticking of a clock beside a bed. When we observe them, or maintain attention to them, they compete successfully for our cognitively controlled attention, and we then have difficulty reading or falling asleep. But over time these noises can cease to attract our observation, and we say that we have "adapted" to them.

An example of maintenance of the considering kind is remembering an event that makes us feel anxious, either because of its effect on us in the past or because it could occur in the future. Such "worries" are too often the targets of our maintenance attention, and because of their motivational significance they are particularly successful in competing with other thoughts for our attention.

The contrast between observing and not observing an ongoing stimulus such a ticking clock or a honking horn may illuminate the processes taking place in visual search and the reading of text. In rapid search and rapid reading, items can apparently be identified without being observed, that is, without producing attentional effects in the anterior cortical areas. When we finally find a targeted item, its successful identification produces the observing event registered in higher-order processes of the prefrontal lobe, and these processes subsequently elevate the perception of the item in location maps of the posterior cortex. During the search process, occasionally a non-target item may be observed in this way. At these moments of observation, the anterior control of attention is dedicated to the item in the posterior location map, and attentional activation of other items is withdrawn.

For the visual skill of rapid reading, on the other hand, an indefinite number of words may be processed without a particular word shape being observed, perhaps because the goal of reading is not the detection of the sensory attributes nor spatial location of a word shape but rather the reception of its meaning that is automatically registered in semantic areas of the brain (LaBerge and Samuels, 1974; Posner and Raichle, 1994). An open problem is to describe the mechanism that produces rapid search and rapid reading in the non-observing mode of processing.

For the case in which memory is scanned for attentional consideration of a topic of current interest, it could be conjectured that searching memories may involve an automatic process similar to the one just discussed for searching visual displays for a target. During the scanning process items in a memory store may be briefly enhanced, but not enough to bring them into attentional consideration. For creative thinking, the activation level of new and novel ideas is presumably quite low relative to the activation levels of familiar ideas, and an understanding of the processes that would bring the weaker creative ideas into full attentional consideration at the appropriate time would be of great benefit.

Explication of the structure and operations of the control, mechanism, and expression of attentional aspects is the main purpose of later chapters, and therefore further detailing of the controlling operations of maintenance attention will not be given here. The point to take from this overview is that the maintenance manifestation of attention shares many properties with the preparatory and selective manifestations of attention, particularly with the sustained property of preparatory attention.

## Summary

In this chapter an attempt has been made to describe preparatory attention as a major manifestation of attention both in daily life and in the laboratory and to introduce the reader to the little-studied topic of maintenance attention. Preparatory attention to an object, location, or action is assumed to be generated when an object or action is expected to occur, but the generation of preparatory attention from an expectancy held in working memory is assumed not to be obligatory but rather under the voluntary control of the individual. Hence, instructions to laboratory subjects to attend to a particular cued location may be retained in memory without an accompanying image of the object in that location being used to build up a high degree of preparatory attention. Yet when the target display appears, the subject may be able to direct attention to the verbally remembered location and respond accurately.

Preparatory attention was described in two general forms: attention to objects, in which imagery plays a particularly important role in generating preparatory attention, and attention to spatial locations, in which the role of orienting was examined. Orienting to exogenous and endogenous cues aligns attention to new sources of information in preparation for processing an object or event there. Orienting in response to sudden onsets of stimuli typically interrupts ongoing attentional processing, even when attention is already directed to the source of information signaled by the orienting cue.

The concept of preparatory attention was used in an attempt to account for attention shifts across the visual field. Several experiments intended to test the moving-spotlight model of attention shifts against the present preparation/selection model were described. Finally, the concept of the allocation of attentional resources was described and examined in light of the increase in knowledge of brain anatomy and function.

Maintenance attention is sustained attention to an object or thought in the absence of any psychological reason other than the intrinsic character of the object or thought itself. A scene may be observed for the pleasure of enjoying its beauty, for example, or a tragic memory may repeatedly come to mind because of the unhappiness it continues to inflict. Although the duration of attention in cases like these may be similar to durations of preparatory attention, maintenance attention does not include the process of expectation served by preparatory attention. It has rarely been the subject of investigation, so a more detailed discussion of this manifestation must await further study.

# 4

## Attentional Processing
## in Cortical Areas

Given the enormous complexity of the nervous system and the abundance of recent research in neuroanatomy, it is obvious that only a small part of what is currently known about brain structure and function can be addressed in the confines of this book. The guiding rule for the selection of research literature to be covered here is to include those anatomical and physiological findings that relate to the expression, mechanisms, and control of attention and to the effects of attention on information flow. The expression of attention as a modulation of information flow in the brain apparently occurs within the circuits of particular areas, and the control exerted upon this attentional expression is presumed to be communicated along relatively long pathways that interconnect the particular areas. Thus, the two neuroarchitectonic levels of local circuitry and systems networks may be viewed as an organizing theme for this section of the book.

Another influence on the choice of neurobiological descriptions of attention stems from the present emphasis on attention in the visual modality, the area in which most of our present knowledge of attention, both behavioral and neurobiological, has arisen. Furthermore, the vast majority of experiments in these two fields of cognitive neuroscience have dealt with perceptual attention, while far fewer studies have been done on attending to actions of either the overt or covert kind. Therefore, the bulk of the neurobiological descriptions to follow will be concerned with several areas of the posterior cortex that apparently specialize in visual perceptual processing, and less descriptive detail will be given for the many areas of the anterior cortex that apparently specialize in the processing of external and internal actions, among other things.

The visual cortex has mainly been studied in the macaque monkey, to a lesser extent in the owl monkey and the human. A few comparisons between the macaque and human may be informative. The surface area of the cortex of one hemisphere of the human brain is about 800 cm$^2$, whereas that of the macaque is about 80–100 cm$^2$ (Blinkov and Glezer, 1968; Van Essen et al., 1992). In both species the major brain regions that make up the visual cortex include most of the occipital lobe plus portions of the posterior parietal and inferior temporal areas (Ungerleider and Mishkin, 1982). It is estimated that approximately 60 percent of the cortex is involved in visual processing for the macaque and from 18 to 31 percent for the human (Van Essen and Maunsell, 1980; Van Essen, 1985). Hence it is not surprising that the majority of research on attention has been concerned with vision.

Three general principles govern the organization of the architecture of the cortex, particularly the posterior cortical visual system. They are: areas of specialization, processing streams, and hierarchical ordering of areas within these streams. These principles of areal organization in the posterior cortex will set the stage for a consideration of how attention modulates the flow of information in visual perception.

## Areas of Specialized (Modular) Processing

The myriad detections, discriminations, and identifications of visual properties of an object or event are a result of various kinds of specialized processing that is relatively automatic. The brain effects this processing not by distributing its computations homogeneously throughout the cortex, or throughout the visual portion of the cortex, but rather by concentrating computations in particular areas for each type of specialized processing. The term *module* is often used to describe these kinds of processing structures. Modular information structures are characterized by being "informationally encapsulated" from the processing going on in other structures of the system and by the "mandatory processing" of input information (Fodor,

1983). The specialized processing of many of the cortical areas indicated in Figure 4.1 are regarded as modular because *specialized* here implies a separation (encapsulation) of a particular process (sequentially or in parallel) from other processes and because, once begun, these processes apparently virtually always run their courses (mandatory processing). In contrast, a distributed processing structure is characterized by interconnections, not separations, between processing elements (e.g., neurons), so that each element participates in virtually all types of processing carried out by the structure. Figure 4.1 (adapted from Andersen et al., 1990; Boussaoud, Desimone, and Ungerleider, 1990; Felleman and Van Essen, 1991) maps out some of the important structures and pathways.

Visual information enters the brain by way of area V1 and the superior colliculus (SC) and flows to areas where specialized pro-

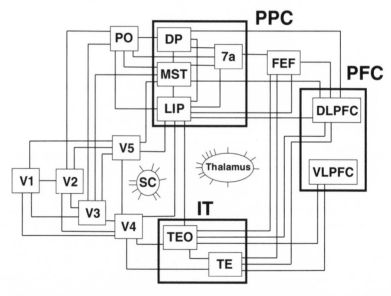

*Figure 4.1.* Cortical (and two subcortical) brain areas involved in visual attention. DLPFC, dorsolateral prefrontal cortex: DP, dorsal prelunate area; FEF, frontal eye fields; IT, inferotemporal cortex; LIP, lateral intraparietal area; PFC, prefrontal cortex; PO, parietal-occipital area; PPC, posterior parietal cortex; VLPFC, ventral lateral prefrontal cortex; and areas MST, TE, TEO, V1–5, and 7a are all parts of the cortex. The superior colliculus (SC) and the thalamus are subcortical structures. (Adapted from Andersen et al. 1990; Boussaoud et al., 1990; Felleman and Van Essen, 1991.)

cessing modules compute various properties of an object or event, such its location and identity. The boxes in the diagram represent a few of the thirty-two or so areas of the visual cortex (Van Essen et al., 1992; Van Essen and Felleman, 1991) in which activity has been shown to be enhanced when a subject is processing some property of an object, such as its location, shape, color, depth, and direction of movement. Of course, at any given moment, visual perception can induce many areas to be highly active simultaneously; when we observe a colorful bird in flight the sensory attributes of color, size, direction and velocity of movement, depth, and shape induce processing in their respective brain areas at the same time. When carefully controlled conditions induce a subject to respond to changes of only one stimulus property, however, activity in a restricted area or set of areas is varied in a preferential manner. Apparently, this localized modulation of activity is possible because the posterior cortex is structured to compute judgments of particular properties of a stimulus object or scene not by a single processor that takes into account the entire stimulus array (in one gulp, as it were), but rather by segregating the input information and directing it to appropriate sets of structures or modules that perform specialized types of computations.

The identification of brain areas that specialize in perception of properties of external stimuli is based on evidence from a variety of techniques, including experiments with subjects who have brain lesions (e.g., Damasio and Damasio, 1989; Shallice, 1988), single-cell recordings (e.g., Allman and Kaas, 1971; Zeki and Shipp, 1988); and positron emission tomography (e.g., Raichle, 1987). These measures are obtained when visual stimuli are presented to the animal in the alert or anesthetized state. Often the neurons in a behaviorally related brain area will show distinctive architectonic properties, such as the presence of heavy myelination or specific transmitters revealed by one of a variety of staining techniques. The characterization of a target area may be aided by observing the pattern of its neural inputs from and projections to other identified areas as revealed by anterograde and retrograde tracing techniques. Finally, a brain area may be convincingly identified if it can be shown that a topographically organized mapping of the contralateral visual field exists within its borders.

The accumulation of evidence that the visual system contains many separable cortical areas beyond the striate projection area of V1 began with the work of Zeki (1969, 1971, 1975) in the macaque monkey and Allman and Kaas (1971, 1974, 1975) in the owl monkey (for a review, see Van Essen, 1985). While Allman and Kaas mapped areas of extrastriate cortex primarily on the basis of the simple visual responsiveness of cells, Zeki found that some areas appear to be specialized in processing particular stimulus attributes of an object, such as its motion or color (Dubner and Zeki, 1971). Although it seemed for a time that definitive processing of a simple visual attribute might be the unique province of a specific brain area, such as color in V4 and motion in V5 (sometimes labeled MT for its location in the medial temporal area of the owl monkey brain, even though in other species it is not located in the same area), further research has led to the view that a series of areas is involved in the processing of a given attribute (Van Essen, 1985; Maunsell and Newsome, 1987). For example, the motion pathways and the color and form pathway appear to be segregated in the blob and inter-blob regions of V1, and flow through V2, with the motion pathway diverging through V3 to the V5 area and the color and form pathway projecting to V4. These two streams may originate earlier than V1 in the magnocellular and parvocellular pathways that flow from the retina through the lateral geniculate nucleus to V1 (Livingston and Hubel, 1984).

Areas of visual specialization (specified by boxes in Figure 4.1) are distinguished by converging data of several kinds, in particular, myeloarchitecture, connectivity, and neuronal response measures. To date, it has been determined that twenty-five areas of the brain are primarily visual, and an additional seven are involved in visually guided motor control and polysensory integrations (Felleman and Van Essen, 1991; Van Essen et al., 1992). Neurotracing techniques applied to the more than three hundred interconnections between these areas have revealed three kinds of projection patterns according to which cortical layers fibers leave or terminate (Rockland and Pandya, 1979; Maunsell and Van Essen, 1983; Felleman and Van Essen, 1991): (1) ascending pathways (forward projections) leave a cortical column from the superficial or both the superficial and deep layers and enter a column at the middle layers, (2) descending pathways (backward projections) leave a column from the superficial or both superficial and deep layers and enter a column at the deep or superficial layers, and (3) lateral (intermediate or same-level projections) leave a column at the superficial and deep layers and enter a column at all layers. The computational significance of these connectivity patterns has yet to be clearly determined, but it has been suggested (e.g., Van Essen et al., 1992) that ascending projections (into middle layers III and IV) are mainly concerned with informational analysis peculiar to that cortical area, while descending pathways are mainly concerned with modulation of the information analyses of that area and with the gating of information flow into and out of the area.

Saying that a brain area specializes in "processing" an attribute is a very general characterization of the kind of computation involved and may conceal important differences between areas. The kinds of computations exerted early in vision, for example, may be different from those occurring in later stages (Van Essen et al., 1992). In early vision, beginning in retinal cells, center-surround antagonisms compute a difference between activity at the center and the mean activity in the immediate surround. This computational operation discards the information represented by the absolute activity values of the center and surround, resulting in a type of selective processing customarily referred to as filtering. Center-surround computations of the filtering kind are apparently repeatedly applied to the flow of visual information as it proceeds through the lateral geniculate nucleus of the thalamus and through the early cortical visual areas as well.

Another type of processing commonly attributed to vision is the "detection event" (e.g., Barlow, 1972; Marr and Hildreth, 1980; Van Essen et al., 1992). This term implies an all-or-none output of the computation, which may be more characteristic of later stages of processing than early stages. An example of an all-or-none detection event is the identification of a color or a face. But prior to that stage of visual processing, the prevailing computation may operate in a filtering manner on the neural signals arising from wavelengths and oriented lines. Thus, it is plausible to regard the specialized processing of an attribute such as color and line orientation as a series of filtering operations taking place within early parts of a cortical stream leading up to a detection operation in a later area.

This book is not an appropriate place to attempt a systematic and detailed review of current knowledge concerning the approximately thirty-two brain areas that participate in visual processing. Rather, we will select for somewhat detailed examination the brain areas that appear to be directly affected by attentional processes. I will describe these areas in more detail as they arise in our considerations of the brain structures that serve the expression, mechanisms, and controls of attention.

*Ventral and dorsal cortical processing streams.* Cortical areas are organized into processing pathways or streams of information flow (Pohl, 1973; Ungerleider and Mishkin, 1982). According to this model, two distinct processing streams arise from area V1; one stream (V1 to V2, V1 to V3, V1 to V4) flows ventrally toward the visual areas of the temporal lobe that are necessary for the discrimination, identification, and recognition of objects, while the other stream (V1 to V5, V1 to V2) flows in a dorsal direction toward the visual areas of the parietal lobe that are necessary for spatial perception and visually guided actions (see Figure 4.1). The original supporting evidence for the dual-pathway idea came from experiments with monkeys with lesions in either the posterior parietal or inferior temporal regions; it was found that parietal lesions impaired spatial discrimination but left intact object discrimination, while inferior temporal lesions impaired object discrimination but left intact spatial discrimination (Ungerleider and Mishkin, 1982; Mishkin et al., 1983).

Early evidence supporting the specialization of shape and location processing in humans came from studies of patients with lesions in the parietal or temporal areas. Lesions in the parietal lobe that impair judgments of location and movement do not impair identification of objects (e.g., Damasio and Benton, 1979), but lesions in the temporal lobe can impair object identification (e.g., Meadows, 1974; Damasio et al., 1992) and discrimination of object attributes of color, brightness, and orientation (Gross et al., 1981; Ungerleider and Mishkin, 1982). More recent evidence for the dual-pathway concept is provided by PET experiments with human subjects that show elevated blood flow in ventral regions of the posterior cortex when subjects perform tasks that involve the discrimination of shape, color, and motion velocity (Corbetta et al., 1991), and elevated blood flow in dorsal regions of the posterior cortex when subjects perform tasks that involve discrimination of spatial location (Corbetta et al., 1993; Haxby et al., 1991). These reports of specialization of dorsal and ventral processing streams do not imply that the streams are separated; anatomical connections exist between the two streams (Felleman and Van Essen, 1991).

The prefrontal cortex also appears to show anatomical specialization for spatial versus pattern processing. Wilson et al. (1993) recorded neuronal activity in two prefrontal areas of monkeys engaged in an oculomotor delayed-response task that required memory for spatial information on some trials and for pattern information on other trials. Neurons in the dorsolateral prefrontal area showed greater activity during delays for the spatial cues than for the pattern cues, and neurons in the inferior prefrontal convexity that is ventrolateral to the dorsolateral area showed greater activity during delays for the pattern cues than for the spatial cues. Some neurons in the inferior convexity responded selectively to faces of monkeys and humans, while others responded selectively to patterns. Jonides et al. (1993) found increased blood flow in right-hemisphere prefrontal and premotor cortices of human subjects while they were engaged in a task involving spatial working memory.

*Hierarchies of areas.* The patterns of interconnectivity between visual cortical areas have been classified in a hierarchical organization of ten levels (Felleman and Van Essen, 1991). Flowing through this hierarchy are streams of information processing, and the brain areas within a stream exert a succession of transformations on the flow of information arising from V1. How one characterizes these streams depends upon whether one looks at the structure or function of the pathways. Cells in the patchy compartments in V1 and stripes in V2 can be anatomically traced back systematically to the parvocellular and magnocellular retinal ganglion cells (Livingston and Hubel, 1984; Zeki and Shipp, 1989; Van Essen et al., 1991). But there is cross-talk between these two pathways (e.g., Schiller et al., 1990), as early as V1 and increasingly so further up the visual-processing hierarchy. If one looks at the pathways within the posterior cortex in the top-down direction, starting with the posterior parietal cortex and the inferotemporal cortex (instead of the V1-upward direction), then one discerns by physiological measurements that the stream joining V1 to the posterior parietal cortex functions to compute spatial properties of the stimulus while the stream joining V1 to the inferotemporal cortex functions to compute the featural properties of the stimulus (for a review, see Desimone and Ungerleider, 1989). Segregating the visual processing of shape and location as the brain does apparently has computational advantages over a scheme in which shape and location are processed within the same module (Rueckl et al., 1989).

A general rule of connectivity between areas is that an area receiving fibers from another area reciprocates the connection. A more specific rule is that the three types of connections between areas (described on pp. 102–103) indicate the position of an area within the hierarchy of processing. A forward projection pattern between area A and area B is almost always combined with a backward projection pattern from area B to area A; this indicates that the processing order in the information flow is from area A to area B. A reciprocating intermediate projection pattern between areas A and B indicates that the two areas are positioned at the same level of the processing hierarchy. It turns out that areas in the early part of the information stream arising from V1 generally exhibit the forward-backward projection pattern (e.g., V1 to V2 to V3), whereas areas in the later part of the stream tend to show the intermediate projection pattern (e.g., V4 to and from V5, and connections between some of the subareas of the posterior parietal area) (Desimone and Ungerleider, 1989). Determining the type of laminar projection pattern between a pair of areas, however, appears to be technically more difficult as the connective distance of the areas from the striate area increases (Boussaoud et al., 1990).

*The expression of attention in brain pathways.* The typical visual scene contains many stimulus objects requiring that the incoming information flow be manipulated or modulated in ways that make the information suitable for effective processing by the appropriate processing modules. For example, when a hiker looks at a dead tree branch to determine whether part of it could serve as a good walking stick, the incoming information from the branch must be manipulated in such a way that only the information arising from a particular attribute or sector of the branch is processed by the module that identifies the class of walking-stick shapes. Similarly, when a reader examines a word that seems to be misspelled in order to find the offending letter, information from just one letter at a time presumably is fed to the appropriate module that identifies single letters.

Reading words and discriminating the attributes of a walking stick are examples of perceptual processes that require selective attention. When searching for a particular word or a particular type of stick, a person may cue herself or himself to prepare to see particular identifying features of that word or that type of stick from ongoing task instructions held in working memory. The attentional preparation for particular features is assumed to take the form of increased levels of activation in posterior cortical maps corresponding to these features (such as line orientations).

It is assumed that both the selective attention process and the preparatory attention process are expressed within the flow of information arising from the visual field as this information courses toward a module that identifies an object. We now ask the more specific question of how the flow of information is modified to deliver the target information to the identifier module or to sustain a preparation to receive featural information of a particular type? But before we can begin to frame a hypothesis in answer to this question, we need to decide the level of information processing that is appropriate for dealing with the modification of information flow between early visual cortical areas (beginning with V1) and the object-identification areas (in the inferotemporal cortex).

One listing of the main levels of organization of the nervous system (based on Shepherd, 1990, and Churchland et al., 1990) gives the following ordering, from largest to smallest: cognitive/behavioral systems, brain areas, neural circuits, neurons, microcircuits, synapses, and molecules and ions. Given these alternative levels of description, the circuit level appears to be the appropriate level of description for the expression of attention, because single-cell recordings (e.g., Chelazzi et al., 1993; Moran and Desimone, 1985; Motter, 1993) and computational considerations (see Chapter 1) show that the expression of attention is well characterized as the difference in activity between certain sets of cells. However, it is not assumed that, at any given moment, selective and preparatory attention are expressed in only one isolated region of the brain, for example in the circuits entering a particular part of the inferotemporal cortex (e.g., during identification of an object). Rather, it is assumed that the expression of selective and preparatory attention may occur simultaneously in many regions of the visual cortex; for example, an object that possesses several attributes could involve selection by location in several cortical maps simultaneously, and cuing a particular set of oriented lines could raise activation levels in several areas simultaneously along the V1-to-IT pathway (see Figure 4.1).

The assumption of the interregional circuit as the appropriate descriptive level for the expression of attention is based on the review of distributed circuits by Goldman-Rakic and her colleagues (Goldman-Rakic, 1988; Goldman-Rakic et al., 1993), in which she points out the extensive reciprocal interconnections between the anterior and posterior cortical areas concerned most directly with location (the dorsolateral prefrontal cortex and the posterior parietal cortex). Furthermore, there are connections between these areas and a broad array of cortical areas (e.g., the supplemental motor area, or SMA, anterior and posterior cingulate area, the parahippocampal gyrus, and the orbital prefrontal cortex) as well as subcortical areas (neostriatum, the substantia nigra and globus pallidus, the superior colliculus, and the thalamus). Thus, when information in the brain arising from a particular location in the visual field is selectively modulated, it is conceivable that corresponding locations coded in the circuitry of many, and sometimes all, of these brain areas are also simultaneously modulated. This distributed view of attentional processing is in accord with the earlier proposal by Mesulam (1981) for a network theory of attention in unilateral neglect (for a discussion of the issue of centralized versus distributed circuitry in spatial attention, see Rizzolatti et al., 1985).

We have provisionally assumed that the circuit level embodies the structures in which attention can be most effectively expressed, and I have tried to show how the choice of this particular level of nervous system activity is strongly influenced by the particular computational definitions of selective and preparatory attention that I have adopted. If selective attention is expressed by a greater flow of information in a target set of fibers relative to other fibers, we then ask what neural mechanism produces this effect. Given that the computational and behavioral goals of preparatory attention are more accurate and faster at processing or reacting to an event in the near future, then it seems reasonable to conjecture that an effective way to achieve these goals is to express preparatory attention at the neurobiological level as a sustained increase in activity in a target set of fibers. But even if preparatory attention were expressed in some other way, we would still be led to the question of what produces the attentional expression in brain pathways. Is the mechanism contained within the circuits that express attention, or is the mechanism located outside these circuits, somewhere it can sculpt the pattern of information flow within the circuits that express attention? The reasoning leading to an answer to this question is presented in the next section of this chapter, in which I review brain areas that have been suggested, by data and computational considerations, as essential or crucial to selective attention.

## Attention to Object Information in Ventral Cortical Streams

In this section I examine areas in the cortex that appear to specialize in performing operations required for preparing for and identifying a visual shape particularly when a distracting object is present. These cortical areas are: the inferotemporal area (IT), the V4 area, the posterior parietal cortex (PPC), and the dorsolateral prefrontal cortex (DLPFC). These regions are shown in Figure 4.1.

While identification of an object apparently requires processing of featural information in area IT, selective attention to the location of an object in a field of distracting objects (e.g., identifying the center object in the stimulus TON as the letter O) is presumed to involve processing of spatial information in the posterior parietal area, where locations of objects are presumably indexed. Furthermore, there must be some way to connect the activities in these two brain regions so that just that featural information that leads to the identification of a particular object is passed to the shape-identification module. This interactive participation of both the spatial and featural streams of information processing in the coarse of object identification is assumed to be necessary not only when the location of a target object is known (i.e., is in working memory, as in the present task) but also when the location of the target object is not known and many available objects are examined successively (as in search tasks). What do we presently know about object identification in the inferotemporal cortex that would be helpful in understanding how attention affects this process?

*The inferotemporal cortex.* The two major parts of the inferotemporal cortex addressed here are the posterior part, which contains a sub-area TEO (near the *te*mporal-*o*ccipital border) that is apparently specialized for the fine discrimination of forms and other attributes of objects (Kikuchi and Iwai, 1980; Spiegler and Mishkin, 1981), and the anterior part of the temporal lobe (sometimes labeled TE), which apparently contains the mnemonic properties necessary for identification of an object (Desimone et al., 1980, 1985). Area TEO lies between V4 and TE. Both parts of IT receive direct projections from V4 (see Figure 4.1), and TE receives inputs from both V4 and TEO (Desimone and Ungerleider, 1989). The familiarity of an object, important to its recognition, appears to involve the superior temporal polysensory area, which is an area adjacent to IT (at least for recognition of faces; Young and Yamane, 1992). Thus, one could make a case that there exist separate areas in the temporal lobe that are crucial for the successful performance of three related but slightly different kinds of behavioral tasks involving visual objects: discrimination tasks, identification tasks, and recognition tasks. Although this chapter is mainly concerned with the effects of selective attention on identification of an object, one should expect that these attentional effects operate similarly for discrimination and recognition of an object.

The kinds of objects to which cells in IT preferentially respond vary widely from simple to complex, including objects made up of various combinations of color and texture with shape (e.g., Gross, Rocha-Miranda et al., 1972; Desimone et al., 1984), faces and hands (Bruce et al., 1981), and toy animals, vegetables, and other natural objects (Tanaka et al., 1991). Variations in location and size of objects do not appreciably change the selectivities of IT cells (Schwartz et al., 1983). Cells in IT also respond selectively to parts of an object, such as faces (Perrett et al., 1982; for reviews, see Perrett et al., 1987; Desimone, 1991).

The finding that some IT cells respond to a whole object and others to its parts is relevant to the way we conceptualize the optional identification of the letter O or the word TON that are embedded in the whole stimulus word STONE. While one part of an object is being identified, other parts are regarded as distractors, and to block the information arising from the distractors from identification processing some selective attentional mechanism is required to operate on the information flow to the object-identification region of IT.

Lesions in monkey posterior IT produce deficits in visual discrimination, and anterior lesions produce deficits in visual identification and recognition (e.g., Cowey and Gross, 1970; Iwai, 1985). Humans with lesions in the IT area experience deficits in recognizing familiar visual objects (Damasio, 1985), and failure to recognize familiar faces (Meadows, 1974; Damasio et al., 1992). Blood-flow studies with human subjects indicate that visual words and pronounceable non-words induce increased activity in the ventral occipital lobe (Petersen et al., 1990).

Receptive fields in the anterior part of IT (TE) are almost always very large, showing a median size of 26 × 26 degrees, and include the center of gaze and portions of the contralateral hemifield (Boussaoud et al., 1991; Gross et al., 1972; Desimone and Gross, 1979). Given the large size of these cells and their overlap with the center of gaze, it is not meaningful to speak of their retinotopic organization, that is, the systematic change in size of receptive field as a function of retinal eccentricity of the stimulus object. For cells in TEO (in the posterior part of IT), however, receptive-field sizes range from a few degrees up to 50 degrees, and their size varies systematically with retinal eccentricity. Therefore, some cells in posterior IT and most cells in anterior IT may respond to an entire computer monitor display.

Information about object identity appears to be vector-coded by a population or ensemble of cells, not uniquely coded or locally coded by one cell (for a comparison of vector and local coding schemes, see Churchland and Sejnowski, 1992). For example, face stimuli appear to evoke responses in a distributed population of cells in IT (Perrett et al., 1987), but while each cell in the ensemble responds somewhat to every face stimulus, particular faces appear to be coded locally in "clumps," that is, by a "sparse ensemble code" (Baylis et al., 1985; Young and Yamane, 1992), in which a relatively few cells may be sufficient to identify a particular object or face. Also within IT are cells that are sensitive to feature ensembles that compose complex shapes as faces and words (e.g., combinations of intersecting bar features that produce L-like, W-like, Y-like, and inverted-T-like patterns). These cells appear to cluster together in groups whose size is estimated to correspond to about 2,000 columns (Fujita et al., 1992). Analysis of the firing patterns of adjacent cells in IT indicate that they carry separate information, as opposed to firing redundantly or interactively (Gawne, et al., 1992).

*Area V4.* Visual information that flows into area IT apparently enters chiefly through the "gateway" of area V4 (Desimone et al., 1990), and this cortical area has yielded the strongest evidence for the modulation effects of attention in the identification of objects (for a review, see Desimone and Ungerleider, 1989). It is not surprising, therefore, that cells in V4 have been found to be sensitive to component features of a visual form. As it turns out, V4 cells compare closely with cells in V1 and V2 in their responsiveness to length, width, orientation, and spatial-frequency features (Desimone and Schein, 1987). Lesions in monkey area V4 produce deficits in form discriminations (Heywood and Cowey, 1987; Schiller and Lee, 1991), and MRI scans of individual human brains show increased event-related potential (ERP) activity in the lateral occipital region during the discrimination of the length of bars whose spatial location had been cued (Mangun et al., 1992).

The major inputs to V4 arise from the early visual areas of V1, V2, and V3, but V4 receives fibers also from a wide variety of other areas, including the temporal, frontal, and parietal cortical areas (for a review, see Felleman and Van Essen, 1991). Of particular importance to selective spatial attention are the projections from posterior parietal areas LIP (lateral intraparietal) and 7a (Seltzer and Pandya, 1980, 1984; Blatt et al., 1990; Anderson et al., 1990), where cells are responsive to the spatial location of a stimulus.

While IT cells typically respond to stimuli that extend well into both hemifields, the cells in V4 respond mainly to stimuli in the contralateral field, and occasionally to stimuli located 1 or 2 degrees into the ipsilateral field (Desimone and Ungerleider, 1989). The receptive-field sizes of V4 cells are generally smaller than those in area IT but often 20–100 times larger in area than those in V1 (Desimone and Schein, 1987). Yet, some V4 cells show peak activity to bars whose width is only 0.05 degree, which is comparable to the discriminative ability of cells in areas V1 and V2 (Desimone and Schein, 1987). The receptive fields of V4 cells have a retinotopic organization, in that neighboring locations on the retina correspond roughly to neighboring cells in V4, and their average size increases linearly from a mean of a few degrees (root receptive field area $= \pi r$) at a retinal eccentricity of zero to approximately 30 degrees at a retinal eccentricity of 50 degrees (Boussaoud et al., 1991; Gattass et al., 1988). In comparison, the receptive fields of V1 cells increase linearly from near zero at zero eccentricity to approximately 4 degrees at retinal eccentricity of 30 degrees (Gattass et al., 1981). Figures 4.2 and 4.3 show the relationship of receptive-field size to the angular

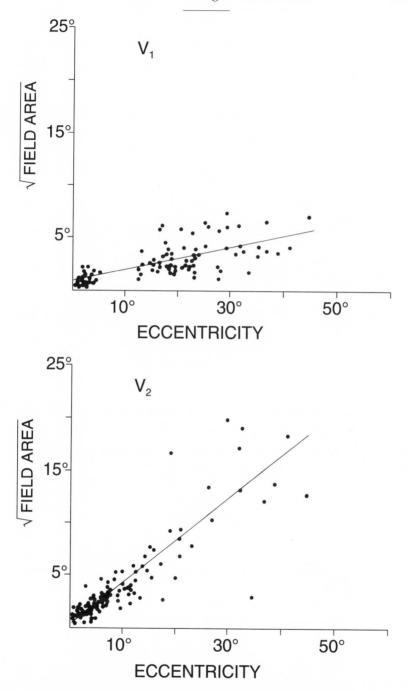

*Figure 4.2.* Receptive-field sizes of cells in visual areas V1 and V2 as a function of eccentricity. (From Gattass, Gross, and Sandell, 1981.)

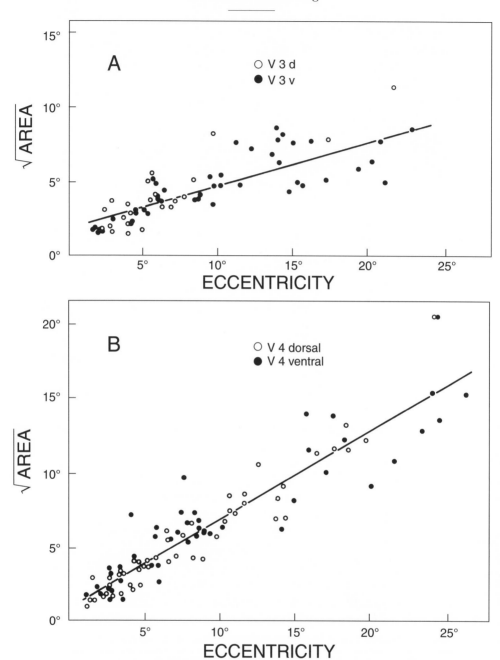

*Figure 4.3.* Receptive-field sizes of cells in visual areas V3 and V4 as a function of eccentricity. (From Gattass, Sousa, and Gross, 1988.)

distance, or eccentricity, from the center of the retina to the center of the receptive field for cells in areas V1, V2, V3, and V4.

The lengths and widths of bar displays that produce preferential responses in V4 cells range from approximately half a degree to 6 degrees of visual arc (Desimone and Schein, 1987). The width of a letter in the displays of our behavioral experiments is about one-third of a degree, and the height approximately two-thirds of a degree, so that the length of a five-letter display such as STONE is almost 2 degrees (including spaces). If one can generalize from the response properties of V4 cells of the monkey to those of the human, it would seem that V4 cells could respond preferentially to the range of sizes represented by the letters and words used in most human behavioral experiments.

Neurons in V4 are of particular interest in this chapter because many of them respond as if they are expressing selective attention. That is, they appear to respond to a stimulus display by modulating information flow at particular locations (Moran and Desimone, 1985; Motter, 1993) and also by modulating information flow corresponding to particular attributes, such as orientation or color (Spitzer et al., 1988, Haenny and Schiller, 1988; Haenny et al., 1988).

Using a matching-to-sample task, Moran and Desimone (1985) trained monkeys to attend to a colored bar shape in one location and to ignore a different one in the other location. When both stimuli were within the receptive field of a cell, the cell responded well if the stimulus (to which the cell was responsive, e.g., a red bar) appeared at the attended location, but it suffered considerable attenuation if the stimulus appeared at the ignored location. Moran and Desimone also examined the case in which the attended location occurred within the receptive field of a cell and the ignored location occurred outside the receptive field of that cell. In this case, the cell's response to the onset of a target stimulus within its receptive field was not contingent on the location at which the monkey directed attention. One could reason that attention to a location outside the receptive field of a cell does not spread inhibitory effects into the field of that cell, and hence there is no contraction of its receptive field nor any attenuation of its response to its preferred stimulus. This effect (confirmed in Chelazzi et al., 1993, but not for all cells in Motter, 1993), which appears to express the selective manifestation of attention, was not found for cells in either V1 or V2.

In another study, Desimone and his associates (Luck et al., 1992) presented stimuli to a monkey one a time, instead of two at a time, and in an alternating fashion between two locations. Their results confirmed the finding that effects of attention at the time of target-stimulus onset are manifest only when both the attended and ignored locations are within the cell's receptive field.

Taken together, the Desimone experiments imply that the expression of attention to a location involves the attenuation of information from the uncued location when it lies within a cell's receptive field, not the enhancement of information flow from the attended location. In other words, the expression of selective attention here involves only a gating (filtering) of information at uncued or unprepared locations and not either a gain (enhancement) alone at the cued location or a gate-gain combination of the two (however, see the discussion of the "suppression versus decay" issue on pages 154 and 193).

Other single-cell studies have shown a gain effect of attentional manipulations on the responding of V4 cells. The tasks in these studies involve presentations of single stimuli that are quite similar to each other, so that subjects are likely to direct relatively high intensities of attention to a specific stimulus. Spitzer et al. (1988), in an experiment that varied the difficulty of an orientation discrimination and a color discrimination in a matching-to-sample task, found that the sample stimulus of the more difficult discrimination produced not only a relative gain (enhancement) in a cell's response but also a narrowing in the tuning curve, suggesting an effect similar to constriction of the receptive field. Haenny et al. (1988) displayed oriented gratings to monkeys in a match-to-sample task and found that the responses of over half of the 192 cells observed in V4 showed a strong response when the monkey had been cued to expect that particular orientation; the cells showed a weak response when the monkey had been cued to expect a different orientation. In a related study, Haenny and Schiller (1988) showed that repetition of a particular oriented grating increased the firing rate of 72 percent of the 154 V4 cells examined and 31 percent of the V1 cells examined. But only the V4 cells showed an additional narrowing of orientation tuning.

Monkeys with lesions of area V4 are severely impaired in selectively responding to the "lesser" of two stimuli, for example the stimulus that is dimmer or smaller or that moves more slowly than the other stimulus in the display (Schiller, 1993). When the "greater" of two presented stimuli is the target, these lesions produce virtually no deficit in performance. In order for the monkey to select a stimulus that induced comparatively less neural activity, it would seem that additional activity must be added to the pathways corresponding to that stimulus. These results suggest that area V4 may function as a "window" through which this additional enhancement is applied to the flow of sensory information. The enhancement mechanism that projects additional activity at target sites in area V4 is assumed in this book to be the thalamocortical circuitry involving the pulvinar nucleus of the thalamus. This mechanism of attention (in its selective, preparatory, and maintenance manifestations) will be described in the next chapter.

*ERP studies of visual and auditory processing during early stages of perception.* The issue of whether attention is expressed in the V1-to-IT pathways as enhancement of activity at the attended object site or as a decrement of activity at the unattended object sites (or both) can be illuminated by ERP measures as well as by single-cell recordings. Hillyard and Mangun (1994) displayed four well-separated squares simultaneously, each square framing a location in a quadrant of the visual field, and subjects were asked to detect a faint target that flashed inside of one of the boxes. Prior to the onset of the target an arrow appeared at the center of the display to cue the square that was most likely to receive the target. On neutral trials four arrows pointed simultaneously to the four squares, and the target subsequently appeared equally often in each square.

The resulting waveforms recorded from scalp electrodes showed larger amplitudes of the positive (P100) and negative (N100) waves at about 100 msec following the onset of the target on validly cued trials than on invalidly cued trials, indicating a modulating effect of attention. Of particular interest here are the separate comparisons of results of the valid and invalid trials with those of the neutral trials. In the case of the P100 wave component, there was no difference in amplitude between the valid and neutral trials, indicating that attention is expressed as a reduction in amplitude on invalidly cued trials, with no apparent enhancement on validly cued trials. For the N100 component, there was a difference in amplitude between valid and neutral trials that was in fact larger than the difference between valid and invalid trials. The invalid trials showed, if anything, a slight increase instead of a decrease in amplitude compared with that of the neutral trials (the finding of an increase in amplitude at both target and surround sites, but more for the target site, has also been found in simulations of thalamic circuit operations described in the next chapter).

Thus, for the P100 component, which maps neural events presumably located along the V1-to-IT pathway, attention appears to be expressed as a decrement of information flow at locations outside the attended area, while for the N100 component, attention appears to be expressed as a relatively strong enhancement of the attended location, combined with a more modest enhancement of the information flow at unattended locations.

Recently, Hillyard and his associates (Heinze et al., in press) carried out a related experiment in which humans matched pairs of letter-like shapes presented to the right and left of center. When subjects were instructed to attend to pairs on one particular side of center, the N100 and P100 waves were enhanced in the contralateral hemisphere. The current dipole that was assumed to generate the P100 wave on the surface field of the brain was localized in the occipital lobe near the border of the temporal lobe. The current dipole is presumed to be the expression of attention at this brain location early in the cortical processing of the letter-like stimuli.

The finding that the attentional modulation of the ERP waveform is mapped onto cortical brain areas known to be involved in early sensory processing receives added support from analyses of the behavioral responses to tasks involving threshold luminance detection. The obtained $d'$ values in the Hillyard and Mangun experiment were 1.59, 1.38, and 1.13, for valid, neutral, and invalid trials, respectively, indicating that these conditions modified processes related to the signal input. Hence the Hillyard and Mangun experiment, together with the Heinze et al. (in press) experiment, provide compelling evidence that attentional operations occur in early visual processing and are not restricted to later stages of decision processing (see also Eimer, 1994).

A similar conclusion with regard to audition appears to be supported particularly strongly by recent studies by Woldorff and Hillyard and their associates (Woldorff and Hillyard, 1991; Woldorff et al., 1994), which address the possibility that the N100 wave enhancement observed in these and other ERP experiments is in fact the summation of two different processes and not the attentional amplification of the dipole generator of the N100 wave. For example, in auditory ERP experiments in which the subject is presented tone sequences to each ear by earphones and instructed to detect an infrequent deviant tone (in pitch or intensity), the obtained enhancement in the N100 wave could be produced from a top-down source of attention that shifts the entire waveform in a negative direction (Naatanen, 1990). There are apparently two kinds of evidence that speak against this superposition hypothesis and favor the specific amplification of the N100 wave. One type of evidence addresses the shape of the N100 wave under attended versus unattended conditions. Under the superposition (or summation) hypothesis, the difference between the N100 wave under the attended and that under unattended conditions should begin before the N100 wave begins to rise toward its peak. The data from one auditory-stream experiment (Woldorff and Hillyard, 1991) shows that, under attended and unattended conditions, the N100 wave begins as the ERP wave rises.

The second kind of evidence supporting the specific amplification hypothesis against the superposition hypothesis is provided by an auditory-stream experiment in which neuromagnetic measurements are combined with standard ERP measurements. Woldorff et al. (1994) used superconducting quantum interference devices (SQUIDs), which consist of thirty-seven magnetic sensors positioned in a circular area of 125 mm diameter placed over the scalp near the left auditory cortex. This neuromagnetic imaging device revealed M100 waves (which are the magnetic counterpart of the electrical N100 waves from ERPs) that were almost identical to the N100 wave shapes. Of particular importance was the finding that the M100 and N100 waves were localized within millimeters of each other in the auditory cortex (on the superotemporal plane). The authors conclude that the M100 generator is the same as the N100 dipole generator located in the primary auditory cortex. Thus, the expression of attention by the dipole generator of the M100 and N100 events, 100 msec or so following the onset of an auditory tone, occurs at the location where the earliest cortical processing of auditory stimuli is known to occur.

Evidence that attention to featural information is expressed in the

V1-to-IT pathway has been provided by PET studies as well as by ERP and single-cell recordings. Blood flow in the ventral occipital areas increases when humans attend to shapes and colors in matching-to-sample tasks (Corbetta et al., 1991), and also when humans engage in a face-matching task (Haxby et al., 1991).

*Some conjectures concerning the enhancement and decrement of attentional activity found in cortical pathways.* The foregoing studies of attention effects in the V1-to-IT pathway, which provide evidence for separate contributions of target enhancement and surround decrement, indicate that attention to an object's location tends to be expressed as an inhibition at the sites surrounding the attended area (Moran and Desimone, 1985; Hillyard and Mangun, 1994) while attention to an object's features tends to be expressed as an enhancement of the target site (Spitzer et al., 1988; Haenny et al., 1988). These findings, taken together, imply that the brain employs more than one class of algorithms to express selective attention. From the data at hand it might appear that spatial attention is expressed by an algorithm that is dominated by surround inhibition, while featural attention is expressed by an algorithm that is dominated by target enhancement. In the experiments cited, however, the spacings of the objects were large and the distinctiveness of the features was low, which suggests that the choice of algorithm may depend instead upon the proximity of the target to non-targets, whether that proximity be spatial or featural (see Motter, 1993, for variations of proximity).

In the cases of well-separated stimuli with relatively easy discrimination demands, as in the Moran and Desimone (1985) task, the location of each object was distinct enough to compete for an eye movement at the onset of the stimulus. Although overt eye movements were inhibited by training, the information corresponding to each eye movement may provide one means for attending to location and ignoring the location when this information is projected to area V4 (Klein, 1980; Rizzolatti et al., 1987). Conceivably, the spatial information appropriate to guiding the eyes to the target and distractor locations is stored briefly in a cortical area such as the posterior parietal cortex, and as information relating to the movement (or goal of the movement) is projected to the superior colliculus, where it is then relayed to the eye-movement motor areas in the brain stem, a copy (or transformed copy) of that information is also sent to area V4. A problem with this account is that long-distance projecting axons are almost always excitatory, so it seems unlikely that the axon fibers projecting from a cortical spatial map to area V4 (and to the superior colliculus, as well) are inhibitory. However, an inhibitory effect in V4 cells could be produced by excitatory afferents that terminate on inhibitory V4 cells.

Another problem with the oculomotor-based account of attention is that some recent evidence (Klein and Pontefract, 1994; Rafal et al., 1989; Wright and Ward, 1994) suggests that while bottom-up (exogenously) controlled shifts of attention are influenced by eye-movement processing, top-down (endogenously) controlled attention shifts are not. The monkey studies cited here cued the location of the target well in advance of the onset of the target, suggesting that attention of the preparatory kind was being controlled top-down from prefrontal areas in which the cue information was presumably being stored. At the present time, it is not clear what selective attentional mechanism is receiving this top-down control and what kind of connection between that mechanism and V4 would account for the observed inhibitory form of the expression of selective attention in V4 cells.

It is possible that when a distractor is positioned sufficiently close to a target (e.g., within a degree), selective attention may be expressed mainly by an enhancement (gain) in firing rate at the target location instead of solely by a decrement (gating or inhibition) of firing rate at the distractor location. When distractors are placed very close to the target there may be no basis for competition of eye movements, if the ensemble is initially registered as one object. Subsequent processing, perhaps of a different type, must align attention around the target area to separate information arising here from information arising from the area of the distractors. Examples of stimulus displays that are presumed to elicit only one eye-movement signal initially because they fit the category of "ensemble perceived as one object" are shown in Figures 3.2 and 3.3 (where target/flanker spacings were on the order of 0.15 degree). In a typical experiment (e.g., LaBerge and Brown, 1989), subjects did not know ahead of time at which of five locations an ensemble triplet (e.g., VRY) would appear. When the ensemble appeared, eye-movement information presumably specified the location of the whole object. At this spacing, increases in target-flanker similarity produce robust increases in human response times; but as the spacing increases beyond a degree, the effects on response time disappear (Eriksen and Eriksen, 1974). Bringing locations of targets and distractors closer may evoke the same kind of attentional enhancement expressed when highly similar orientations produce confusions in discrimination tasks (e.g., Spitzer et al., 1988). When locations of targets and distractors are close, the featural information in these locations is more likely to interact and produce confusions; consequently, successful selective attention may require more than simply an attenuation of the surround, namely the additional sharpening properties that lie in enhancements of the target area.

Thus the mechanism that produces selective attentional expression mainly by variations in enhancements (gains) at the target location may be different from the mechanism that produces it mainly by variations in attenuations (filtering, gating) at the surrounding locations. In each case, however, if the expression of selective visual attention by spatial location takes place at or near V4 in the V1-to-IT pathway, there must exist appropriate anatomical connections between each mechanism and this area.

## Attention to Spatial Information in Dorsal Cortical Streams

The notion of space used in the disciplines of cognitive science appears to denote not only location but also extent, that is, not only the "where" of something but also the "size" of that something. For example, in giving a complete spatial description of an object, such as a large rock near my left foot, the system not only computes its location in the visual field by a point representing its "center of gravity"; it also computes the important spatial property of its apparent size, which could range from that of a small pebble to a large boulder. This dual description of spatial information is particularly significant when the spatial computation by the system is modulated by attentional operations. A cat that sees a rodent disappear under some leaves is paying attention to where the mouse disappeared but presumably also to whether the rodent is the size of a small mouse or a large rat. Furthermore, the size of spatial attention determines whether a part of an object, the whole object, or a cluster of objects is being processed in terms of some (featural) attribute; for example, a cat's attention may be momentarily concentrated on a group of small mice, a particular mouse, or on the sharp teeth located on one end of a particular mouse. A guiding question of this chapter concerns what attention does to the processing of spatial information in cortical areas of the brain which affects the perceptual judgments of objects and the performance of bodily responses in space.

Attention to space has been emphasized before in this book in reference to the identification of objects when other objects are displayed in the visual field. The identification of friend or foe, suitable food, written words on the page, clothes to wear, objects to buy in the marketplace—all seem to require that information be selected according to its specific location (and size). The simple spatial registration of an array of displayed objects on the part of a computational system is only the beginning of a hierarchical chain of operations that results in the identification of an object or a group of objects. Following the topographical registration of objects, there must be a way to index a particular object or cluster of objects for specialized processing by a task-designated judgment module (a module that performs identification, or discrimination, or even an assessment of aesthetic value). Attentional operations presumably provide the required modulation of the information projecting from the registration of the stimulus array to the judgment module so that only the information arising from a particular location (and size) is passed to the judgment module.

Another way, and a particularly important way, that locational information is used is as a means of marking or indexing the point in environmental space where a response may be directed, and attentional operations improve performance of this function by modulating the processing of marked locations both in space and time. Examples of the necessity to mark a location in space prior to the performance of a response are reaching, grasping, and locomotion over a rough terrain. Spatial information seems much less important to the generation of other familiar responses, such as talking and singing, and a salient difference between these two classes of responses is that in the former case the overt movements themselves are related in a direct way to the spatial properties of objects; the direction and extent of hand movements in reaching and grasping and of leg movements in walking are contingent on where nearby objects are situated. In talking and singing, however, the overt movements are normally unconstrained by the spatial arrangements of

objects in the immediate environment; for example, bodily movements a person uses to speak or sing about a beautiful flower are
not ordinarily affected by the location or size of the flower. But for
many spatially related movements it is not at all unusual that their
success depends on very precise computations of object locations:
consider the dexterity required to click a computer mouse on a small
target on the video monitor, finger a chord on a violin, or hit a
racquetball. The adaptive performance of a great many kinds of perceptual judgments and motor responses requires the quick and accurate computation of spatial information linked to the timely execution of an appropriate response.

The ultimate neuroscience goal here is to describe how these
kinds of spatial computations are modulated by attentional operations in neural tissue. I begin the survey of neural structures involved
in spatial attention with the posterior parietal cortex, which is the
brain area that has been most frequently implicated in the literature
devoted to spatial attention.

*Posterior parietal cortex.* Attentional selection of an object for identification in the temporal lobe may occur by means of several attributes, such as color or orientation. Another and perhaps more frequent way that selection of an object takes place is by means of its
location in the visual field or its position within a larger object or
group of objects. Connected directly to the V3–V5–V4–TEO–TE
areas is a posterior parietal cortex (PPC) area called the lateral
intraparietal (LIP) area, which specializes in the processing of
spatial information underlying locational and positional selection
(for a review of LIP connections see Blatt et al., 1990). The PPC
is reciprocally connected with the dorsolateral prefrontal cortex
(DLPFC), which appears to function as a working memory for spatial
information (Goldman-Rakic, 1987). The major subcortical areas
specializing in visual spatial functions are the superior colliculus,
which computes information that is relatively directly concerned
with eye movements (e.g., Schiller, 1984; Sparks, 1986), and the thalamus (e.g., Jones, 1985; Steriade et al., 1990). The possible roles of
these two subcortical structures in spatial attention will be described
in the next chapter. Meanwhile, it is well to keep in mind that the
functioning of the PPC, superior colliculus, and thalamus should
not be treated in relative isolation individually or even as a closed
group, because these structures are interconnected extensively with
other brain areas as well as with each other.

Evidence that the PPC is involved in spatial processing, and in particular spatial attentional processing, comes from a variety of experimental methodologies and measures. These include: human lesion studies (e.g., Critchley, 1953; Heilman et al., 1985; Posner et al., 1984) and monkey lesion studies (e.g., Lynch and McClaren, 1989; Stein, 1978) showing that attention may not be redirected to the contralateral field; single-cell recordings in monkeys (e.g., Motter and Mountcastle, 1981; Mountcastle et al., 1975; Bushnell et al., 1981; Andersen et al., 1985; Robinson et al., 1978; Hyvarinen, 1982; Goldberg et al., 1990) that show increased firing in PPC cells during spatial tasks; ERP studies with humans, carried out with MRI scans, that show increased negativity over the PPC 150–190 msec after a visual event occurred (Mangun, et al., 1992); cerebral blood-flow studies indicating increased activation in the superior parietal cortex during spatial vision tasks (Haxby et al., 1991) and during a spatial cuing task (Corbetta et al., 1993).

The PPC is divided into two major subareas (for a review, see Andersen et al., 1990): the superior lobule contains somatosensory area 5 and the inferior lobule contains subareas positioned around the intraparietal sulcus, including visual area 7a, somatosensory area 7b, visual lateral intraparietal area (LIP), and dorsal prelunate area (DP). Additional areas included in this grouping are: an area on the medial surface of the cortex, labeled 7m (Goldman-Rakic, 1987; Goldman-Rakic et al., 1993), and the superior temporal area (MST) concerned with motion. In the human, the superior lobule contains both areas 5 and 7 (Von Bonin and Bailey, 1947), and the inferior lobule includes the supramarginal gyrus (area 39) and the angular gyrus (area 40).

The major cortical connections to the PPC revealed by neurotracing studies that relate most directly to selective attentional processing may be divided into three groups: those originating in V1 and ascending the hierarchy of levels, those connecting to V4 and the temporal areas, and those descending from prefrontal areas. Virtually all connections can be regarded as reciprocal: projections ascending from V1 originate from superficial and deep layers and terminate in the middle layers; projections descending from the DLPFC originate and terminate in superficial and deep layers; and lateral projections (between areas at the same hierarchical level) originate in the superficial and deep layers and terminate in all layers (Andersen et al., 1990; Felleman and Van Essen, 1991).

The first group of connections to the monkey PPC considered here enables flow from V1 through intermediate areas of the visual hierarchy (see Figure 4.1) to enter the inferior parietal areas (DP, LIP, 7a, and MST) from areas V3, V4, V5, and PO (the parietal-occipital area, sometimes labeled V6). These latter areas, in turn, receive feedback (descending) projections from certain areas of PPC that potentially could modulate ascending visual-information flow in an attentional manner. Specifically, PPC visual areas LIP and DP have direct descending projections to area V4, the attentional modulation effects on which were discussed in the preceding section of this chapter.

The second group of connections project laterally from LIP to TEO and from 7a to IT, to STS (superior temporal sulcus), and the parahippocampal gyrus (Andersen et al., 1990). Conceivably, these connections could enable spatial information about an object in PPC to modulate featural information of an object in these higher processing areas in the temporal lobe by constricting the effective receptive fields of cells in the manner hypothesized by Moran and Desimone (1985) for cells in V4.

*Coding spatial location in the posterior parietal cortex.* Receptive field size of a cell provides an indication of the way a representation such as spatial location is coded in a brain area such as the PPC. Within the PPC is area LIP, which contains cells that are light-sensitive, saccade-related, and memory-related (Gnadt and Andersen, 1988; Goldberg et al., 1990), and area 7a does also (Andersen et al., 1990). In one study (Blatt et al., 1990) receptive field sizes for light-sensitive cells in 7a were found to be very large and bilateral (extending 15 to 25 degrees on each side of the vertical meridian), while receptive field sizes for light-sensitive cells in LIP ranged from a few degrees near the fovea to about 15 × 15 degrees at about 25 degrees from the fovea; most receptive fields were contralateral to the stimulus location. Unlike the fields of IT and V4 neurons, receptive fields in LIP and 7a typically do not overlap the fovea. However, the slope of the line relating receptive field size to eccentricity was similar to that of V3 and V4 (slope = 0.30, Gattass et al., 1988); the intercept for LIP was about 11 degrees while that of the line for V3 and V4 was 1–2 degrees (see Figure 4.3). Thus, area LIP shows a course topological representation of visual space peripheral to the fovea. Moreover, the arrangement of cells in the LIP area showed a rough topographical relationship to the location of their receptive fields in external space. Cells in 7a show a much larger receptive field size and therefore indicate an even coarser coding of locations in external space.

Visually related and eye-movement related cells in LIP and in 7a change their firing rates linearly as a function of horizontal and vertical eye position (Andersen et al., 1985; Andersen, 1987; Andersen et al., 1990), indicating that information from both eye position and retinal position are combined in these areas to yield a head-centered coordinate representation of space. These computations were successfully simulated in a neural network that took points in retinal coordinates as inputs and as output the frequency of firing, which is a scalar code that is appropriate for driving mechanisms that control eye saccades (Zipser and Andersen, 1988; Andersen and Zipser, 1988; Goodman and Andersen, 1990). The layer of units in the network representing the input-output transformation was modeled as a distributed processing code, which suggests that a distributed representation of external space in the PPC could produce outputs that exhibit a spatial accuracy similar to that of internally generated eye movements.

When one talks of the "sector" of the circuitry of a brain area that carries spatial information, one is tempted to envision a "real" map in which topographical relationships of the cells in brain space correspond to the topographical relationships of external space (roughly speaking, the ordering of points is preserved between cells and external space). This kind of relationship exists for the lateral geniculate nucleus (LGN) and V1, even though the metric relationship of locations is distorted in these maps to give more neural representation to the foveal sector of the retina. In some (but not all, see Andersen, 1989) of the posterior parietal regions of the monkey, however, visual location information appears to be distributed rather than ordered in topographical correspondence with locations in the visual field.

Distributing spatial information does not imply, of course, that spatial information arising from the external visual field is lost. What is lost is the ability to compute in a local fashion in these areas phenomena that may be based on spatial distance, such as the width of a selected attentional area of visual space and, of particular interest here, a target area/surround selection effect involving some kind of lateral inhibition. As one proceeds from area V1 to areas further along the several parallel streams emanating from V1, the size of the receptive field of a typical cell increases, and the information corresponding to a given location in the visual field is carried by more and more cells. The precision of representation of visual space may, in principle, be recovered by the combined processing of an ensemble of such cells, where each cell coarsely codes a spatial location (Hinton et al., 1986; O'Reilly et al., 1990) in these advanced visual-processing areas.

Additional support for the claim that highly precise spatial locations can be coded by coarse-grained inputs is given by Hinton et al. (1986) and Ballard (1986), who describe how the number and degree of receptive-field overlap are related to precision of location coding in the network. More general treatments of distributed activity in neuronal populations are given by Churchland and Sejnowski (1992), Freeman (1975), and Lehky and Sejnowski (1990).

O'Reilly et al. (1990) used distributed-network simulations to compare the accuracy of spatial encoding by units resembling cells in PPC and IT. Receptive-field peaks in IT tend to be located at the fovea (Gross et al., 1972), while those in 7a and LIP are spread out and tend to avoid the fovea (Blatt et al., 1990). O'Reilly and co-workers found that a parallel-distributed-processing network mapped a point of a retinal input with less error as the number of off-fovea receptive-field peaks increased, favoring the PPC cell-types over the IT cell-types in spatial mapping accuracy by more than a factor of 10 (when the number of off-fovea receptive-field peaks reached 24). Therefore, this study joins the simulation studies of Andersen and his colleagues in showing why, computationally, areas of the PPC may be well designed for representing locations in space. (For similar kinds of analyses of the computational advantages of separating the "what" and "where" processing streams instead of combining them within one module, see Rueckl et al., 1989.)

Thus there are two major views of how spatial-information processing related to locational and positional selection may be represented in a brain area, such as the PPC or DLPFC. They are the topological and distributed views, and these different types of representation explain in different ways how attentional operations might modulate information flow in and around a restricted spatial area of a particular size. The topological view, perhaps the more traditional one, regards the PPC as containing relatively explicit maps of body-centered space in which each cell is finely tuned to a point in visual space (i.e., coding in the PPC is local). The cells in a representation can be finely tuned to external space and yet may, unlike retinal receptor cells, themselves be scrambled within their anatomical area. The cost of scrambling anatomical positions of cells that function topographically is the increase in the lengths of interconnecting fibers. The distributed view, which appears to be gaining wider acceptance recently, regards the PPC as containing relatively implicit maps in which each cell is broadly (coarsely) tuned to many points in visual space, so that, in effect, the resulting representation of a point is distributed across many or all the cells within the representation (i.e., vector coding). It is almost always the case that each element in a vector-coded representation is coarsely coded or tuned. In a distributed network the unit of representation is the strength of an individual synapse, but in the topographical network the unit of representation is the single neuron.

Some investigators regard the distributed view as particularly suitable to the purpose of coordinating sensorimotor functions, such as moving the eyes or limbs toward a visual target, because, in principle, sensorimotor functions require only an appropriate transformation of vectors in retinotopic coordinates into a vector in motor coordinates. On the other hand, the topological view may be regarded as particularly suitable to the purpose of efficiently coordinating more cognitive functions, such as generalization, attention, and search (Stein, 1992; Robinson, 1992).

It is important to note that a particular representation may be realized through a coding scheme that is intermediate between finely tuned topographical coding and completely distributed coding. For example (as discussed in the IT section, above), a subset (cluster or clump) of cells may respond strongly to a stimulus while other cells respond moderately, and still others not at all (for a more detailed discussion of these distinctions in coding schemes, see Churchland and Sejnowski, 1992). Moreover, over an interval of 100–200 msec, the location of the active cell cluster may move across the representational field, as in the example of the effect of stimulating one whisker of the rat upon clusters of cells in the thalamus (Nicolelis et al., 1993); another example is eye-movement control in the superior colliculus (Muñoz and Wurtz, 1992).

We turn now to the potential relationship of cell responses in the PPC to the operation of spatial attention. The studies of Andersen and his associates (e.g., Andersen et al., 1990) and Goldberg and his associates (e.g., Goldberg et al., 1990) amply demonstrate that cells in the PPC can code the information needed to move the eyes to a particular location in space. The former group of investigators apparently favors the view that these cells code an intended movement, while the latter apparently favors the view that these cells code the spatial location of the target of the movement in oculomotor coordinates. The question related to selective attention is whether or not this information could also be utilized to select a particular spatial location by known projection routes from the PPC that terminate in areas V3, V4, or TEO. Cells in the PPC in monkeys respond selectively not only when the animals intend to move the eyes toward an object but also when they intend to move a limb toward an object to touch it (Bushnell et al., 1981). The topography of eye movements is different from that of reaching, and yet the same cells in the PPC apparently can control both kinds of movements. In addition, brain connectivity strongly suggests that the spatial information in a stimulus reaching the DLPFC and the frontal eye fields (FEF) is relayed through the PPC; the output of the PPC network of spatial representation to these frontal areas could be organized differently than outputs for eye movements or limb movements.

In view of these considerations, we inquire into the possibility that the same cells in the PPC that represent spatial information that is relayed to the FEF and DLPFC working-memory areas, and to the areas that control eye movements in the SC and reaching movements in the anterior cortex motor areas, might also project spatial information to cells in the V3–V4–TEO pathway that are assumed to express selective attention, even if the topography of spatial coding in this pathway differs from that within the PPC. The fact that the topographical coding is less coarse in V4 and TEO than, say, in LIP need not prevent an accurate copying of spatial information from LIP to V4 and TEO, since the connections are reciprocal and the required precision of mapping between these areas could be learned during the early developmental history of the individual.

*The prefrontal cortex.* The third group of connections link the PPC with the prefrontal visual spatial areas 8 (another name for the FEF, frontal eye fields) and 46 (also called the principal sulcus, PS) (Blatt et al., 1990; Andersen et al., 1990; Cavada and Goldman-Rakic, 1989), where cells respond to and store location information of visual objects (Bruce et al., 1985; Funahashi et al., 1989, 1990; Goldman-Rakic, 1987). Reciprocal projections exist between area 7a and the prefrontal areas 8 and 46 (Pandya et al., 1971), but the stronger connections are between 7a and 46 (Andersen et al., 1990). Apparently there is a topographical correspondence between parts of LIP and parts of prefrontal cortices 8 (FEF) and 46 (Blatt et al., 1990).

The prefrontal areas, important to the internal voluntary control of spatial attention, project to the PPC areas in a descending fashion from the prefrontal areas. Goldman-Rakic and her associates (1993) measured activity of cells in both the prefrontal and posterior parietal areas of an animal engaged in performing an oculomotor delayed-response (ODR) task. This task requires the animal to move the eyes to the location of a visual target that had been briefly flashed several seconds earlier. Single-cell recordings were taken from the prefrontal and posterior cortices of the animal. Histograms of discharges from pairs of cells, one in the prefrontal cortex and the other in the PPC, showed highly similar profiles; some pairs of cells responded to the location of the cue while it was present, other pairs responded to the direction of the upcoming eye movement during a three-second delay period, and other pairs responded to the direction of the eye movement when it occurred. An additional measurement of activity in the prefrontal and PPC areas was provided by a comparative analysis of local cerebral glucose utilization (LCGU) for two groups of monkeys, one of which performed a delayed spatial-response task (involving working memory and therefore prefrontal activity) and the other a visual-pattern discrimination (involving associative memory and presumably much less prefrontal activity). The results showed that the working-memory task increased LCGU about 16 percent across the dorsal and ventral banks of area 46, and by the same amount in PPC areas (7a, 7b, 7m, and the intraparietal sulcus that contains LIP).

Recently, PET measures of blood flow while humans subjects performed a spatial-location task showed increases in prefrontal, occipital, parietal, and premotor areas of the cortex, but only in the right hemisphere (Jonides et al., 1993). The involvement of many cortical areas in spatial memory is consistent with the distributed-processing view of working memory proposed by Goldman-Rakic (1987), in which the prefrontal cortex has a dominant role in controlling activity in other areas. It could be conjectured that much of the activity in these other cortical areas are expressions of attention, while the control of these attentional expressions resides with activity patterns of the prefrontal cortex.

While spatial control appears to be centralized in the dorsolateral part of the prefrontal cortex (DLPFC), recent evidence points to the centralized control of object identity in the inferior convexity, which is ventrally adjacent to the DLPFC and linked to object areas in IT (Bates et al., 1994). After giving monkeys a delayed-response task that involved color or shape information in identifying an object, Wilson et al. (1993) found delay-sensitive neurons for object identity in the inferior convexity area. These authors interpreted their findings as indicating that the inferior convexity records the identity of an object over a time delay in much the same way that the DLPFC records the location of an object over a time delay.

These results strongly suggest that storage of the location of an object over a short period of time (between the time of an instructive locational cue and the signal to respond, say, or even over a block of trials) occurs in prefrontal areas, and that cells in spatial maps here can project this coded spatial information to corresponding PPC spatial maps in two important ways: (1) in a tonic (sustained) mode, as required, for example, to maintain preparatory attention to a location, or to maintain a bodily posture in readiness to make a sensorimotor response toward an object, or (2) in a phasic (brief) mode, as required, for example, to narrow spatial attention to quickly select one object in a cluttered field, or to trigger a sensorimotor response to an object.

The conjectured relationship of the connectivities of the PPC and PFC areas to selective visual attention would fit the following scenario: It appears that spatial information is required computationally in the V1-to-IT pathway to restrict the featural information entering identification modules in IT to just the information arising from a particular object, for example, identifying the center letter in the display STONE. Since the choice of the center letter is entirely arbitrary (one could instruct the subject to identify the last letter instead), some means external to the V1-to-IT pathway must modulate the information flow to allow only the information from the center letter to enter an identification module. Instructions as to which item to identify in an ensemble such as a five-letter word must typically be stored for at least a short period of time in working memory, and then at the time the stimulus display appears these instructions must be converted into the operation of selective filtering or enhancement (or both) of the featural information in the V1-to-IT pathway. Evidence is strong that the storage of information in working memory takes place in the prefrontal area, and it seems reasonable to hypothesize that utilization of that stored information takes place either via projections from the PFC directly to the V1-to-IT pathway or indirectly to the V1-to-IT pathway through the PPC.

The input to PPC that selects a location is presumed to originate either from external visual stimulation (via the ascending pathways from V1) or from internal sources, such as working-memory areas that store spatial instructions (Goldman-Rakic, 1988). For example, when a monkey chooses one foodwell over another (as in a delayed spatial-response task), spatial attention presumably operates to process the information (spatial and/or featural information) arising from the location of that foodwell, while the information arising from the other foodwell location is attenuated. In order to challenge the selective-attention mechanism strongly in this situation, the foodwells may have to be positioned closer together than is customary in laboratory experiments. One may reason, following the instructed identification of the center letter of a word in humans, that if a monkey is cued to attend to the location of a particular foodwell during a delay, the location of that foodwell is presumably stored in the DLPFC network and is either continuously copied to appropriate networks in the PPC (activating memory cells there) or copied to the PPC at the time that the animal is cued to respond.

Given the lack of knowledge of cell behavior in homologous structures in the human, there may be problems with generalizing this scheme to humans engaged in identifying the center letter of a letter string such as a word, particularly when the string appears at or near the fovea. Since there is no clear evidence, to my knowledge, that monkey PPC cells can effectively code fine-grained locations at the foveal center (although Colby et al., 1993, shows PPC cells whose receptive fields overlap the fovea), then one should exercise caution when generalizing from monkey to human cells with respect to representing spatial information at the foveal region. One would expect, however, that the ability to discriminate locations at or near the center of the visual field has adaptive value for the monkey (e.g., in picking small ants from a stick or detecting small parasites in fur), but the adaptive value may not approach that of the human that can detect letter locations that typically subtend an angle of a third of a degree. When human subjects identify the letter B within the string BRK (against BPK and BQK), the error rate is less than 10 percent even when these strings appear 1.8 degrees away from foveal center (the most sensitive region of the eye). Observing the BRK string here on the page (at the typical viewing distance of reading) confirms that the eye can distinctly move to each of the three locations when the string is at the fovea, and even when the string appears a few degrees to one side of the foveal center a distinctive saccade can still be guided to each letter. Therefore, it would seem that saccadic movements may be directed across the small locational differences corresponding to the position of each letter in the BRK string.

Since the BRK string may be flashed in an experiment for 217 msec, there is insufficient time to move the eyes to sustain the stimulus at the foveal center before it goes off, but attention can apparently shift to the location of the string and select the area of the center letter before the stimulus goes off. Therefore, it seems reasonable to conjecture that the representation of spatial information in the brain that guides a potential eye movement to the location of the R in BRK could also guide the shift of attention to its location. Given the apparently large receptive fields in the PPC, however, the coarse coding of locations of closely spaced objects would seem to require sharpening. The brain structure that is directly connected to the PPC and apparently has the requisite circuitry for sharpening distributions of activity in cortical maps is the thalamus. In the next chapter I consider the converging evidence for the hypothesis that the thalamus is involved in selective attention, particularly when the spatial locations of objects are close and when their features are highly similar.

*Anterior cingulate area.* On the medial surface of the frontal cortex is an area that has been implicated in the control or management of task operations (see Figure 4.4). An example of operations management occurs in the multiple matching-to-sample task employed by Corbetta et al. (1991) in a PET experiment. In this study, human

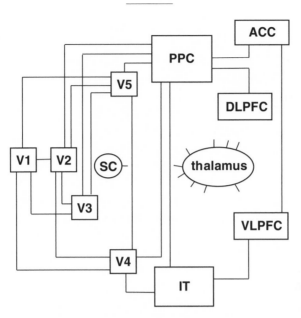

*Figure 4.4.* Anterior and posterior cortical areas involved in visual attention: the anterior cingulate cortex (ACC), dorsolateral prefrontal cortex (DLPFC), and ventrolateral prefrontal cortex (VLPFC).

subjects were briefly shown two successive displays of moving colored rectangles and were asked to judge whether the displays differed with respect to either movement velocity, color, or shape of rectangle. The most difficult processing was required for sequentially applying the particular operations involved in judging color, shape, and velocity of movement the moment that a display appeared. A contrasting condition merely required the subjects to make one type of judgment (either color, or shape, or velocity of movement) of a display in a block of trials. Given this simpler condition there was no problem of operations management on a trial because the one type of operation could quickly become routinized after a few trials.

The PET measures taken during the complex color-shape-movement task showed increased blood flow in the right anterior cingulate cortical area (ACC) and in the right dorsolateral prefrontal area (DLPFC); these two cortical areas did not show increased blood flow during task conditions in which only one attribute (color, or shape, or movement) was judged.

Other tasks that have shown increased blood flow in the anterior cingulate area measured by PET are verb-generation tasks (Petersen et al., 1988), in which a subject is shown a noun and then generates an associated verb (e.g., *hammer* → *hit*), and the Stroop task (Pardo et al., 1990), in which a subject is asked to name the color of ink in which color words are printed. Both of these tasks appear to confront the subject with a set of competing operations, which induces the subject to construct a suitable routine that manages these operations. During early practice in performing these kinds of tasks subjects apparently direct attention to specific operations, but after considerable practice subjects can perform even these complex tasks in a routine manner that requires much less attention to the management of individual operations. Hence one would expect that the management computations of the anterior cingulate would be less evident in PET measures during later stages of learning, and recent experiments have borne out this expectation (Raichle et al., in press).

## Attentional Control versus Attentional Expression

The distinction between control and expression of attention, emphasized strongly in this book, can be used here to help explain the differences in the ways that cortical brain areas operate during attentional processing. According to the present view, the anterior cortex contains areas that are crucial to working memory for object locations, object attributes, and the operations that may be performed on this stored information. These specialized areas of working memory are assumed also to induce and modulate activity in other areas of the cortex, both anterior and posterior areas, that process information concerned with object location and object attributes as well as operations. Thus, the working-memory areas are said to control the expressions of attention that occur in areas concerned with information processing other than the temporary storage of information.

A contrasting view (Shiffrin and Schneider, 1977) does not distinguish control and expression of attention in this way, but rather confines attentional processing to operations within short-term memory structures. Thus, attention is assumed to be involved in the movement of information into short-term memory from sensory memory, and in the movement of information to and from long-term memory, and in the movement of information within the short-term store.

When a subject is asked to perform two or more tasks at a time, the subject must engage in the management or coordination of attention to the particular goals and operations of each task at each moment. It has been proposed that these "executive" or organizational activities are processed by areas in the frontal lobes (Baddeley, 1986, 1992; Norman and Shallice, 1980; Duncan, 1994). Most complex problems (e.g., problems in standard intelligence tests) can be analyzed into subproblems, each of whose solutions may be defined in terms of a goal, and a particular subproblem may be further analyzed into its subproblems, whose solutions may again be defined in terms of a goal. In this way the entire problem may be conceptualized and stored as a hierarchy of goals (Duncan, 1994). After an individual has solved the problem and gone through it several times, each remembered subgoal may itself index in memory the operations (internal or external actions) that are needed to achieve it.

While the individual is engaged in organized problem-solving, attention is involved in the selection of the appropriate subgoal from an array of other subgoals, as well as in the selection of the operation that will achieve that subgoal. Then, attention must be sustained for a time to an ongoing operation in order to protect that operation from interfering information from other concurrent operations. For example, in the coordination of two hands in playing the piano or in opening a door while holding grocery bags, the control of operations is assumed to be shifted rapidly between the two hands, and during the time intervals that signals are being sent to change the operations in one hand, the information must be insulated so that interfering cross-talk does not arise from other ongoing operations (Allport, 1989).

Goals and operations are presumed to be processed in pathways of both posterior and anterior cortex, and the expression of attention in these pathways is assumed to be the relative enhancement of information flow in the pathways corresponding to the attended goal or operation relative to the information flow in the pathways corresponding to the unattended goals or operations. This conceptualization of attentional selection in the organizational management of executive actions is the same as the conceptualization of attentional selection in the modulation of incoming sensory information discussed in earlier chapters. Examples of areas that express task goals involving locations and attributes of objects are in the posterior parietal and inferotemporal areas, among others areas. Example of areas that express task operations are the supplemental motor areas, motor areas, and related somatosensory areas (e.g., Roland, 1985, 1993; see also Boschert et al., 1983; Brunia, 1993; Hackley et al., 1990).

What is being emphasized here is that areas that express attention may not be the same areas that control this expression. The controlling areas under present discussion are assumed to contain a working-memory component and a voluntary-based component that can allow the contents of working memory to activate attentional expressions in their corresponding areas of the cortex. This "two-process" assumption of working memory was described earlier in the book (Chapter 3) in connection with the distinction between expectation and preparatory attention, where expectations were assumed to be stored representations in working memory that could (at the person's option) be used to induce an expression of preparatory attention in perceptual pathways.

The notion that working memory for object locations and object attributes is specialized in the prefrontal areas of the DLPFC and the inferior convexity has been supported by PET and single-cell recordings during tasks that fairly well specify the delay properties that are measured by these laboratory techniques. However, the hypothesis that the working memory for operations or procedures is specialized in the anterior cingulate is at present less well supported by experiments, partly because the tasks used in experiments so far do not separate the working memory for operations from the management of these operations. Since operations are almost always temporarily stored during their management, it is not clear whether the anterior cingulate area shows involvement because of working memory or operations management or both.

Until the accumulation of further data clarifies the issue, it may not be unreasonable to assume that the function of the anterior cingulate area in performing the kinds of complex tasks described here is one of holding operations in working memory while the processes that "manage" operations are carried out in other closely connected areas, such as the supplemental motor area; attention to specific operations during the initial learning of a complex task would then be expressed in these areas. The content of the memory item in the anterior cingulate area may be substantially more complex than items typically held in spatial or object memories of the DLPFC and inferior complexity areas. A remembered procedure may contain several operations, and when an operation is activated a "place marker" may be required to index how much of the operation has been performed and how much is yet to be performed. Thus, the complexity of accessing as well as storing sets of operations would seem to require an order of complexity that is considerably higher than is presently known for storing object location and object attributes.

## Summary

This chapter has extracted from the burgeoning evidence arising from neuroscience laboratories the information most relevant to understanding how attention may be expressed and controlled in cortical brain pathways. Several conclusions appear to be supported by these findings. (1) Attention may be expressed in many different cortical areas simultaneously, as happens when a subject is attending to a particular object (e.g., in V4) and attending to its location (in DLPFC and PPC). (2) The expression of attention in a brain area appears to be described effectively as an enhancement of activity in the attended set of pathways relative to the activity in the unattended set of pathways. (3) Anterior cortical areas that serve working memory for object locations, object attributes, and operations on object representations are assumed to control the expression of attention in the cortical areas in which attention is expressed as object locations, attributes, and operations are being processed.

# 5

## Attentional Processing in Two Subcortical Areas

The two subcortical structures that have been most strongly implicated as mechanisms that modulate attentional processing are the superior colliculus and the thalamus. Another subcortical structure, the basal ganglia, exerts an indirect effect on attention in cortical pathways by means of its projections to the superior colliculus and to some nuclei of the thalamus. In addition, a variety of subcortical neuromodulatory nuclei affect attentional activity in a diffuse manner through the release of neuroactive substances in the cortex, thalamus, and superior colliculus. This chapter will be mainly concerned with the involvement of the superior colliculus and thalamus in visual attention.

It should be made clear at the outset that attentional computations in one structure need not be directly communicated to a cortical area in order to affect the expression of attention there. An important example is the superior colliculus, whose output axons synapse in the pulvinar nucleus of the thalamus en route to cortical areas. Another example is the prefrontal cortical areas, which are involved in spatial working memory and which send axons to the posterior parietal "maps"; these in turn project axons to areas in the V1-to-IT pathway (e.g., V4), where selective attention is expressed to an object's location.

### The Superior Colliculus

The momentary orientation of the eye provides the individual with only a part of the total available visual information. Likewise, the

140

momentary position of the hand on a surface provides the individual with only a part of the total tactual information available from the immediate environment. In contrast, the ear can pick up auditory information from all directions. Fine adjustments in orienting—the alignment of the sensitive fovea of the eye or the sensitive tips of the fingers with a sector of the incoming information—can dramatically improve certain kinds of information arriving at the eye and hand. For the ear as well, fine adjustments in head orientation (and/or pinnae in the case of animals that can effectively move the ears) toward a sound source can increase the individual's sensitivity to sound. Thus the process of orienting enables the system to gain advantageous access to particular sources of information in the environment.

A guiding theme in the present description of the superior colliculus is orientation: how the superior colliculus may employ orienting as a mechanism of attention, and how the outputs of this mechanism may generate an expression of attention in cortical pathways. Related to issues of expression and mechanism is the matter of the control of orientation-based attention, in particular the control by activity arising from stimuli in the external world, termed exogenous control, and control from plans within the individual, termed endogenous control.

One of the most influential theories of attention today is based on the notion of orienting (Posner, 1980, 1994). Posner distinguishes overt orienting, which involves movement of the eyes, from covert orienting, which shifts attention from one location to another in the visual field in the absence of eye movements (e.g., Eriksen and Hoffman, 1972; Posner, Nissen, and Ogden, 1978). Moreover, it appears that prior to an eye movement attention is quickly shifted to the new location in the visual field and serves to guide the trajectory of the eye movement to the target location. The process of shifting attention is assumed to involve three distinguishable components: disengaging attention from a current location, movement of attention to the new location, and the engagement of attention at the new location.

The three components of attention shifting in Posner's orienting theory are captured elegantly in the spatial cuing task developed by him. Three boxes, each approximately 1 degree in size, are displayed on a visual screen, one at the center and the others 8 degrees to the left or right of center. One of the three boxes is brightened for 150 msec, after which (at variable time intervals) an asterisk is presented in one of the three boxes. The subjects are instructed to respond with a key press when the asterisk appears. To identify the three components of attention shifts, the investigator may choose examples from many different pairs of trial events in this experiment. One example is a trial in which the left square is brightened and the asterisk is presented in the right square. The attention shift of interest here is from the left square to the right square. The first component of the shift is the disengagement of attention from the location of the left square, the second component is the movement of attention from the left to the right square locations, and the third component is the engagement of attention at the location of the right square or, more specifically, at the location of the asterisk within the right square.

The movement component of attention shifts is assumed to involve a network within the superior colliculus, and the disengagement and engagement components are assumed to involve networks in the posterior parietal lobe and pulvinar nucleus, respectively. A movement component of attention shifts involving circuits of the superior colliculus would presumably influence the form with which attention is expressed in cortical pathways, for example, in area V4 and in the posterior parietal areas. The concept of attentional expression in cortical pathways emphasized in this book can be framed within the orienting theory of attention by extrapolating from the engagement component of attention shifts. When attention becomes engaged by virtue of pulvinar operations, the output of the pulvinar network onto cortical pathways becomes the expression of attention. The form of the attentional expression is presumably shaped not only by the pulvinar network that projects directly to cortical pathways but also by the superior colliculus network, which projects its activity to cortical areas through synapses within pulvinar circuits (see Figure 5.1, below).

If the same circuit mechanisms that move the eyes also move attention, then it will be helpful to examine the structure and functions of these mechanisms. Although the superior colliculus is emphasized in this section of the chapter, other brain structures, such as the frontal eye fields (FEF) and the posterior parietal areas (PPC), also exert strong influences on eye movements and must be considered together with the superior colliculus in accounting for eye movements and covert orienting of attention.

*Structure and connections to other brain areas.* Of the many brain areas that can participate in the function of orienting the eye, the superior colliculus shares with the FEF the privileged role of issuing commands to the oculomotor nucleus, which innervates the muscles that move the fovea to a new location (Schiller, 1984; Sparks and Pollack, 1977). Movement of the fovea can be accomplished by moving the eyes, the head, and the trunk of the body. In addition, the superior colliculus contains maps of visual, auditory, and somatosensory space that overlap and are aligned with each other and are in register with motor maps (Meredith and Stein, 1990); multisensory inputs from the location of an environmental object can therefore be effectively integrated to promote fast and efficient responses to the object.

The superior colliculus is a "hill" located on the dorsal surface (the "roof" or tectum) of the midbrain, one structure on each side of the brain midline. It has an oblong shape, and in the human its size is approximately 6 mm along the anterior-posterior axis, 8 mm in the medial-lateral axis, and 4 mm in the dorsal-ventral axis (Watson, 1985). In the cat the corresponding measurements are 5, 5, and 3 mm, respectively (Meredith and Stein, 1990). As these dimensions imply, the human colliculus represents a relatively small phylogenetic increase in volume over that of the cat, compared with the substantially large increases in the volumes of the pulvinar and the cortex (Jones, 1985) from the cat to the human. Volume ratios of the superior colliculus and the lateral geniculate nucleus (LGN) in the hamster, rat, and rhesus monkey are 3:1, 2:1, and 1:8, respectively (Schiller, 1984).

The superior colliculus is a highly laminated structure made up of seven alternating cellular and fibrous layers (Schiller, 1984; Sparks, 1986). The seven layers are classified anatomically into three superficial, two intermediate, and two deep groups but are often classified functionally into three superficial layers and four deep groups. Figure 5.1 shows the superior colliculus partitioned into two areas, the superficial ($SC_s$) and the deep ($SC_d$) areas.

*The superficial layers of the superior colliculus.* The superficial layers receive almost all of their inputs from the contralateral retina and from the ipsilateral visual cortex (Sparks, 1986), and cells in these layers discharge in response to a visual stimulus. About 10 percent of retinal ganglion cells project to these layers (Van Essen et al., 1992), and the percentage of cells that show increased firing rates

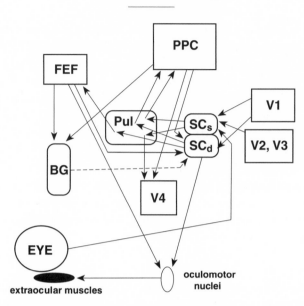

*Figure 5.1.* Major connections of the superior colliculus (represented here by its superficial, $SC_s$, and deep, $SC_d$, layers) with cortical areas, the pulvinar nucleus of the thalamus, the basal ganglia (BG), and the oculomotor nuclei that innervate the extraocular eye muscles. Solid lines indicate excitatory, and the dashed line inhibitory, connections. (Based on Schiller, 1984; Sparks, 1986.)

increases as one probes deeper into these layers (Wurtz and Mohler, 1976). From the superficial layers visual information is directed through synapses in the inferior and lateral pulvinar to many cortical areas (Abramson and Chalupa, 1988). The two major cortical targets of the superior colliculus are the posterior parietal area (Benevento and Fallon, 1975; Bender, 1981; for reviews see Chalupa, 1991, and Huerta and Harting, 1984), and areas within the V1-to-TEO pathway (Benevento and Rezak, 1976; LeVay and Gilbert, 1976). Thus, sensory information arising from the retina reaches the posterior cortex not only by way of the LGN and striate cortex but also through the superior colliculus.

Receptive fields of superficial cells (in the cat) show a mean diameter of $16.4 \pm 12.1$ degrees, with a range of 2 to 81 degrees (Meredith and Stein, 1990), while those of deep cells responding only to visual stimuli show a mean diameter of $48.5 \pm 26.7$ degrees (cells that respond to auditory and somatosensory stimuli as well as visual show mean receptive-field diameters that range from 72 to 86 degrees). These large receptive fields are not surprising in view of the fact that the retinal input to the superior colliculus is by way of Y-cells and W-like cells, which themselves exhibit large receptive fields (Marracco and Li, 1977; Schiller and Malpeli, 1977). The superficial layers contain a retinotopic map of the contralateral visual field (for a review see Meredith and Stein, 1990).

*Deep layers of the superior colliculus.* The deep layer of the superior colliculus receives inputs from parietal cortex (Lynch et al., 1985), the visual cortex (Fries, 1984), and the frontal eye fields (Stanton et al., 1988), as well as from the subcortical basal ganglia. Cells in the deep layers of the SC are organized in topographical maps representing not only visual (Stein and Arigbede, 1972; Meredith and Stein, 1990) but also auditory (King and Palmer, 1985) and somatosensory modalities (Stein et al 1976). The relatively large size of the receptive fields of deep-layer cells suggests that this structure may signal the direction and amplitude of saccadic eye movements by vector (ensemble) coding (for a review see McIlwain, 1991).

Unlike the superficial area, which receives strong cortical inputs from V1, V2, and V3, the deep layers in the monkey SC receive their main cortical visual inputs from the posterior parietal area (Lynch et al., 1985), from prefrontal areas (Goldman and Nauta, 1976), and the frontal eye fields (Leichnetz et al., 1981). The deep layers contain a map of visual space that is stacked adjacent to maps from auditory and somatosensory spaces in a manner that the cells responding to corresponding points in space lie along the same vertical axis (Meredith and Stein, 1990). Stimulation of these cells by microelectrodes produces movements of the eyes and head (Schiller, 1970; Hikosaka and Wurtz, 1983) and the ears as well (Meredith and Stein, 1985); these movements are controlled by projections to nuclei in the brainstem and spinal cord. In particular, cells in the deep layers show an increase in firing rate just prior to a saccadic eye movement (Goldberg and Wurtz, 1970; Schiller and Koerner, 1971), whether the saccade is made in the light or in the dark.

*The influence of the basal ganglia on the superior colliculus.* The projections from the basal ganglia to the SC (Graybiel and Ragsdale, 1979) are of particular importance because they tonically inhibit activity in the SC cells (see Figure 5.1). The specific part of the basal ganglia that contains the cells that project to the SC is the substantia nigra pars reticulata (SNr). It is well known that these cells are inhibitory (that is, their axon fibers secrete the neurotransmitter gamma-aminobutyric acid, or GABA) and therefore inhibit activity of SC cells (Hikosaka and Wurtz, 1985). It is also known that these inhibitory axons terminate mainly in the intermediate layers, less frequently in the deep layers, and rarely in the superficial layers (Hikosaka and Wurtz, 1983). Since the cells in the intermediate and deep layers project to the generators of eye movements in the brainstem, it would appear that their functional role is the tonic inhibition of commands to these brainstem eye-movement generators. Cells in the SNr cease firing just prior to the onset of eye movements, indicating that the SNr influences the initiation of eye movements by disinhibition. The neural events that inhibit cells in SNr can be traced backward along the inhibitory projections from the caudate nucleus to the SNr, and the excitatory projections can be traced from all areas of the cortex to the caudate nucleus.

The neural connections relevant to the generation of an eye-movement signal from the intermediate and deep layers of the SC are not restricted to projections from the SNr, since both the PPC (Lynch et al., 1985) and the FEF (Illing and Graybiel, 1985) project to the intermediate layers. Therefore it seems possible that the SC signals an eye movement by combining the excitatory influences from the FEF and PPC with the inhibitory and disinhibitory influences from the SNr (Hikosaka and Wurtz, 1989).

*Control of eye movement.* The two main types of eye movements are pursuit movements and saccadic movements. A pursuit movement is the relatively slow and continuous displacement of the eyes in their orbits that maintains the alignment of the fovea with a target object while the head or target (or both) is moving. A saccadic movement brings an object into alignment with the fovea by a fast ballistic displacement of the eye in its orbit.

The muscular control of an eye in its orbit involves six extraocular muscles that are innervated by the third, fourth, and six cranial nerves, whose soma are located in the brainstem. The components of the vector of oculomotor control are direction and amplitude. The direction of an eye movement is controlled by the relative amounts of activity in the six muscles and the size (or angle of displacement) of the eye in its orbit is controlled by the amplitude of firing. Studies (Robinson, 1970; Schiller, 1970) have shown that the relationship between displacement in orbit and firing rate is generally linear.

Cells in the intermediate and deep layers of the superior colliculus control the brainstem oculomotor neurons (which also control the orienting movements of the head, neck, and pinnae). Selected cells here discharge immediately before a saccade, with the frequency of firing increasing rapidly up to the onset of the saccadic movement, and then there is a sudden termination of firing (Schiller and Koerner, 1971; Wurtz and Goldberg, 1972a). However, other eye-movement cells here show a loose coupling between discharge rate and eye movement (Sparks et al., 1976; Mays and Sparks, 1980), suggesting that the triggering event for the eye movement arises from some other source, perhaps a releasing of the substantia nigra tonic inhibition of the superior collicular cells, since these cells are gated by inhibitory input from the SNr (Hikosaka and Wurtz, 1983; Karabelas and Moschovakis, 1985). Nevertheless, prior to all eye movements, cells in the intermediate and/or deep layers of the superior colliculus discharge along with the motor neurons in the brainstem (Wurtz and Goldberg, 1972a).

Cells in the superficial layers of the superior colliculus are responsive to direct visual input from the retina (from about 10 percent of ganglion cells) and the visual cortex, but their activity does not necessarily result in an eye movement. When an eye movement does follow a visual stimulus, however, it appears to be one of two general types: a normal saccade or an express saccade. Normal saccades occur on the order of 150–250 msec following stimulus onset, while express saccades can occur as fast as 80 msec following stimulus onset (Schiller, Sandell, and Maunsell, 1987; Kingstone and Klein, 1993; Fischer and Weber, 1993). Express saccades typically occur when there is no stimulus present at the current fixation (at the center of the visual field), while normal saccades are made when a stimulus is present. Express saccades leave very little time for computations elsewhere than in the superior colliculus, while normal saccades may leave as much as 100 msec for computations to be performed in other brain regions (Schiller, 1984; Sparks and Nelson, 1987).

Recently Muñoz and Wurtz (1993a,b) have located cells in the

rostral pole of the monkey superior colliculus (deep layers) that increase their activity when the monkeys are engaged in fixating a visual target in the center (+ 1–3 degrees) of the visual field. During
saccades, these cells showed a pause in their discharge rate. When
muscimol was injected into the rostral area, resulting in an inhibition of the fixation cells, express-like saccades occurred at latencies
less than 100 msec.

When individuals shift attention without moving the eyes, many
of the eye-movement computations may take place while the actual
triggering of the eye movement is blocked. That is, cells in the superficial layer may register a retinotopically coded visual event, and cells
in the intermediate and deep layers may compute an eye-movement
vector (direction and amplitude), but these computations may not
produce an eye movement. This state of affairs may be explained
by noting that the superior colliculus is subject to modulation by
projections from other brain areas (e.g., Wurtz et al., 1980; Wallace
et al., 1991). In particular, projections to the (monkey) superficial
layer arise from cortical areas 17, 18, and 19, the frontal eye fields,
the middle temporal area, and the posterior parietal area (Fries,
1984; Graham et al., 1979). Cortical projections to the deeper layers
originate in the frontal eye fields (Leichnetz et al., 1981), the prefrontal cortex (Goldman and Nauta, 1976), and the posterior parietal area (Lynch et al., 1985).

In addition to the connections of the SC with other brain structures already mentioned, commissural connections are also relevant
to the movement of the eyes. It is a general anatomical principle
that most muscles are arranged as opposing pairs, one agonistic and
the other antagonistic. The signals that control these muscles, therefore, must be coordinated if an effective movement is to result
from their combined action. All of the main controls on eye movements upstream from the brainstem structures have direct commissural connections that can serve as a basis for coordination of eye-
movement control and therefore potentially represent the opposing
kinds of signals ultimately sent from the SC to the brainstem generators of eye movements. These possible representations in brain structures of the excitatory and inhibitory effects that give rise to an eye
movement or a potential eye movement may have important implications for our understanding of the mechanisms that produce the
expression of attention in brain pathways. At the same time, the pos-

sibility that these representations may be distributed across major cortical and subcortical structures suggests that the algorithms involved in such a mechanism could be quite complicated.

While cells in the deep layer of the SC project to the brainstem nuclei that drive the muscles of the eye, cells there also project via the thalamus to the posterior parietal areas and frontal eye fields (Harting et al., 1980; for a review see Sparks, 1986), whose cells show enhanced firing rates when saccadic eye movements occur (Goldberg and Bushnell, 1980; Mountcastle et al., 1975).

Given the proximity of the superficial, intermediate, and deep layers within the SC, it is somewhat surprising that the existence of functional visual interconnections between these layers is controversial. For a recent review of evidence against functional interconnections between the superficial and deep layers the reader is referred to Meredith and Stein (1990), and for another review and evidence favoring the functional interconnection hypothesis the reader is referred to Behan and Appell (1992). This debate probably remains unresolved because the circuitry of the superior colliculus is somewhat complicated (Sparks, 1986). The circuitry of the thalamus, on the other hand, consists of a relatively simple repeating-modular structure, and as a result questions about input-output connectivities and interactions between sectors of the thalamus can be addressed in a relatively clear manner.

This brief review of the connections between the superior colliculus and both the posterior parietal area and the frontal eye fields, which themselves are strongly interconnected, suggests that the coding of spatial information in these cortical areas may be in a form that is directly useful for the generation of eye movements by the superior colliculus. This view is supported by the fact that the superior colliculus serves as an alternative route from the retina to these cortical areas that is phylogenetically older and more reflexive in character than the geniculostriate route. The retinal cells giving rise to information in this route are the Y-cells, which also project to the same areas via the geniculostriate pathway. Thus the Y-cell reception of the location of rapid-onset stimuli is redundantly projected through these two major pathways. While the cells in the superior colliculus respond to visual, auditory, and somatosensory inputs, they respond preferentially to the spatial location of the stimulus.

*A brief review of attentional effects found in the superior colliculus.* The first experiment that provided evidence for attention-like effects in the SC was a study of monkeys by Goldberg and Wurtz (1972a), who found cells in the superficial layers whose firing rates would change depending on whether or not the monkey was about to move its eyes toward a visual target. In one task, the monkey was trained to fixate on a point of light at the center of the screen and to press a lever when the light began to dim, meanwhile ignoring a peripheral light. In another task, the monkey was required to move the eyes toward the peripheral stimulus and to press a lever when it dimmed. Cells that responded for about 400 msec when the peripheral stimulus was in their receptive fields significantly increased their firing immediately following the onset of the stimulus and before the eye movement began. The activity of these cells was apparently closely related to the stimulus and not to the evocation of the response, since the cells did not respond to eye movements made in the absence of a visual target. Furthermore, when the target of an eye movement fell outside the receptive field of one of these cells, the cell did not show the substantial increase in presaccadic activity exhibited when the target falls within the receptive field. This marked increase in firing rate of a cell in response to a visual stimulus was called "enhancement" by Goldberg and Wurtz, and they proposed that it served as a substrate of the phenomenon of visual attention.

Subsequent experiments showed a similar presaccadic enhancement effect in the PPC (Bushnell et al., 1981), the frontal eye fields (Goldberg and Bushnell, 1981), the extrastriate cortex (Fischer and Boch, 1981), and the pulvinar (Petersen et al., 1985). The enhancement effect was found to be independent of an eye movement in the PPC (Bushnell et al., 1981) when the animal was required to respond to the dimming of a peripheral stimulus in some way that did not involve saccadic movements or to respond by reaching to touch the stimulus with the hand. An early study found no enhancement effect in the SC, however, when the monkey attended to the target stimulus without moving the eye to it (Wurtz and Mohler, 1976), but a more recent study that used a matching-to-sample task revealed enhancement of cell firing in the SC in response to a stimulus in the absence of eye movements (Gattass and Desimone, 1991). In this task, the test display contained both a target (to be matched to the preceding cue) and a distractor; SC superficial-layer cells showed enhanced firing when the attended target stimulus was inside their receptive fields but showed no enhancement when an un-attended distractor was inside their receptive fields. One possible explanation for the difference in results may be that the matching-to-sample task produces more intense attentional preparation at the attended location than is elicited by the Wurtz and Mohler (1976) peripheral dimming task.

Goldberg and his colleagues maintained that the earlier findings indicated that the enhancement effect found in the SC is not a sub-strate for visuospatial attention (Goldberg and Colby, 1989). Even when the enhancement effect in some tasks depends upon an eye movement, however, the enhancement of cell firings in the SC could still be regarded as an instance of attentional processing. In the task used by Goldberg and Wurtz (1972) and Wurtz and Mohler (1976), the movement of the eye toward the peripheral stimulus was the trained response to be made in the presence of two stimuli, the fixa-tion point and the peripheral stimulus point. Although the monkey is attending to the center fixation point before the peripheral stimu-lus is added to the display, the onset of the peripheral stimulus pro-duces a display containing a target and a distractor in which the animal has been trained now to regard the peripheral stimulus as the target and the fixation stimulus as the distractor. When the pe-ripheral stimulus appears, the selective attention mechanism pro-jects an appropriate activity pattern across the SC cells that will result in an eye movement to the peripheral stimulus instead of mainte-nance of the eye position at the fixation point. Presumably this out-put from the selection mechanism reaches other areas as well, not only the intermediate and deep areas that signal the generation of eye movements in brainstem nuclei but also the areas in the PPC and the frontal lobe that represent locations of the stimuli in space and the basal ganglia areas that influence the release of inhibition on SC cells.

Apparently some cells in the intermediate layer of the SC respond to a visual stimulus by increasing and maintaining activity for several seconds in the absence of a subsequent eye movement (Mays and Sparks, 1980). Named ''quasi-visual'' (QV) cells, these cells appear to encode a planned eye movement in a manner similar to that of lateral intraparietal (LIP) cells (Gnadt and Andersen, 1988). Given the direct projections to the intermediate layer from the PPC and FEF, it would seem plausible to conjecture that these QV cells in the

SC are a part of the system responsible for the distributed expression of attention that includes the PPC and parts of the V1-to-IT pathway; and, through their participation in the expression of attention, these cells come under the control of frontal areas concerned with spatial memory (FEF and DLPFC).

The responses of monkey SC cells were recorded in a pair of recent studies. Gattass and Desimone (1991, 1992) used a matching-to-sample task in which a distracting stimulus was presented along with the test stimulus and found that cells responded better when the stimulus in the receptive field was the attended stimulus than when it was the distractor. When electrical stimulation was directed to superficial-layer cells at the location of the distractor, performance was impaired, but when electrical stimulation was directed at the location of the target (when the distractor was either present or absent), there was no impairment of performance. These results suggest that SC cells can be modulated by the attentional conditions of a task and that the modulatory effects seem to arise from cortical areas.

Human patients with lesions of the superior colliculus (and proximal tectal areas), diagnosed as having supranuclear ophthalmoplegia, show abnormalities in orienting, particularly in the vertical direction. Rafal et al. (1988) presented a spatial orienting task (Posner, 1980) to a group of these patients and found that they showed slower attention shifts in the vertical direction than in the horizontal direction. The control group of patients with Parkinson's disease showed no difference in vertical and horizontal attention shifts. The authors interpreted the findings as indicating that midbrain (SC) structures influence the movement component of shifting attention from one location to another in the visual field.

*The superior colliculus as a possible mechanism that produces an expression of attention in cortical pathways.* In view of the anatomical connections of the SC to other brain areas and the response properties of SC cells, it seems plausible to conjecture at this time that the SC serves to express attention by the enhancement of cell firings during those intervals when eye movements are not being made. Sometimes these enhancements are brief, as in the Wurtz and Goldberg (1972a) experiment, and sometimes they are several seconds in duration, as in the responses of the QV cells of the Mays and Sparks (1980) experiment. If, as the Gattass and Desimone (1991, 1992) experiments suggest, the enhancement of firing of SC cells in attention experiments can be produced from cortical sources, then it seems plausible to hypothesize that the SC together with these same cortical sources could project the same enhancements to areas in the V1-to-IT pathway—for example, V4, where selective attentional effects have been demonstrated.

What is the expected form of the expression of attention in the V1-to-IT pathway produced by activity projected from the parietal areas and/or from the SC? The fibers from the SC synapse first in the pulvinar, where they are subject to the type of enhancement to be described below, in the discussion of the thalamus. Some fibers from the parietal areas project directly to the V1-to-IT pathway (Felleman and Van Essen, 1991), while others appear to project through pulvinar pathways. What seems clear is that all of these projections are excitatory (as a rule, long-distance projections in the brain are excitatory; the main exceptions are outputs from the basal ganglia). Thus, it would be expected that attention-related mechanisms in the SC and parietal areas would enhance cells in the V1-to-IT pathway.

If circuit mechanisms of attention project only excitatory activity to the expression of attention in the V1-to-IT pathway, how does one account for the data from single-cell recordings in the V1-to-IT pathway of monkeys engaged in spatial attention tasks in which attention appears to be expressed by some cells as an inhibition or suppression of the unattended distractor site (Moran and Desimone, 1985; Motter, 1993)? Two explanations are suggested here. One explanation assumes that some of the long-distance excitatory fibers from the parietal and pulvinar cells fall on inhibitory cells within the V1-to-IT pathway, thus inducing local inhibition. Thus, the parietal and pulvinar projections fall on both excitatory and inhibitory cells, and the amount of activity corresponding to each depends upon whether a specific location is a target or a distractor. A second explanation assumes that parietal and pulvinar axons fall predominantly on excitatory cells in the V1-to-IT pathway; that at the onset of the stimulus display both target and distractor locations are enhanced. Subsequently, when one particular site (the attended site) receives added activation from attentional controls in prefrontal areas (that project through parietal areas), activation is sustained in corresponding sites of SC and V4 maps while activation decays at the other site (the unattended site). The reduction in activation at distractor sites therefore occurs not by active suppression but by passive decay. Thus the rate of cell firings at a non-attended site would rise at first and then decay, while the rate of cell firings for an attended site would rise to some level and be sustained there while the period of attention lasts.

*Control of orienting mechanisms.* How might conjectured SC mechanisms of attention described in the foregoing section be controlled? The separation of covert visual-orienting controls into external or exogenous sources and internal or endogenous sources was spearheaded by experiments of Jonides (1980, 1981). The theoretical assumption that the same general oculomotor mechanism was being controlled by exogenous and endogenous sources was incorporated in the oculomotor readiness hypothesis by Klein (1980) and more recently formulated in the "pre-motor" theory of Rizzolatti et al. (1987; see also Sheliga et al., 1994). The part of the hypothesis that assumes exogenous control of oculomotor (SC) mechanisms during covert visual orienting seems widely accepted, but there currently exists some controversy concerning whether endogenous voluntary control of orienting involves the oculomotor mechanisms (e.g., Reuter-Lorenz and Fendrich, 1992; McCormick and Klein, 1990; Rafal et al., 1989; Wright and Ward, 1993).

*Contrasting mechanisms of orienting with mechanisms of enhancement and selection.* When attention has been conceptualized within the context of visual orienting, it has taken on many of the properties that are exhibited by the movement of the eyes. Hence the term *covert orienting* has become almost synonymous with attention, particularly in situations requiring a shift of attention. As the distance of the attention shift begins to exceed a degree or so (beyond the area in which SC fixation cells operate), then eye-movement mechanisms are presumed to be strongly activated, particularly by sudden onsets of stimuli, and the mechanisms will produce an actual movement unless the subject employs internal controls to prevent them from occurring. In these "long-distance shift" tasks, oculomotor mechanisms may be involved during the shifting of attention but other kinds of mechanisms may be required after the shift has occurred to filter out the distracting information from other stimuli. For example, the target of an orienting response may be a group of objects, such as three people standing near each other, or a variety of foods on a plate, or several letters in a word. After attention has been shifted to the people, plate, or word, subsequent attentional operations of selection may be required to identify components of those groupings, and those attentional operations may involve operations that are different from those involved in orienting.

Figure 5.2 illustrates the distinction being drawn here between attentional shifting and attentional selection. The trial begins with a solitary plus sign at the center of a display, followed by the presentation of two letters to the right or to the left of center. For this example, the duration of the presentation may be long enough to permit eye movements to each letter, or it may too brief to allow an eye movement. When the letters are well separated, the retinal position of each letter is sufficiently separated to allow clear resolution of the information arising from the attended letter. When the letters are clustered together, however, the attention shift occurs to the group of letters, and neither overt nor covert eye movements can resolve the information arising from one of the letters. Even if the pair of letters falls at the center of the retina, it will register virtually equal quality of information from each letter regardless of where the eye is positioned (assuming, of course, that the distance between the eyes and the display is not unusually short).

Mechanisms of orienting, therefore, would seem to serve attention mainly when a target object is sufficiently separated from other objects. The amount of separation that calls upon operations of covert orienting may correspond to the amount of separation that normally evokes an eye movement to bring the target to retinal center. But

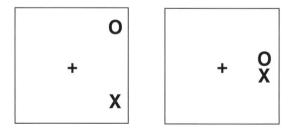

*Figure 5.2.* Selection of target information by simple orienting *(left display)* or by the addition of attentional resolution *(right display).* The plus sign is presented first alone, followed by the presentation of the two letters to the right or left of the plus sign.

even in some cases of short eye movements or short shifts of attention from one object to another, the information arising from two objects may overlap sufficiently to require additional resolution. The crucial assumption being made here is that the information arising from each of the clustered letters cannot be separated by the same orientation-based mechanism that separates the information arising from the distant locations of the two letters in the other display. Some other mechanism or mechanisms would seem to be required to filter out distracting information. A central hypothesis of this book is that the filtering process is carried out by the circuitry of the thalamus. The structure and function of the thalamic circuitry in filtering or selecting information is the subject of the next section of this chapter.

It should be pointed out that simply presenting distractors adjacent to a target does not guarantee that spatial filtering operations are required in order to resolve the information arising from the target. The other important requirement is that the distractor be similar to the target (the property of similarity includes similar direction and velocity of movement, if the displayed objects are not stationary). Figure 5.3 illustrates the case in which the target and the distractor are similar, and can be compared with Figure 5.2, in which the target and distractor are dissimilar. If target and distractors are sufficiently dissimilar, then the target information arising from the primary visual area may be separated by virtue of the difference in attribute information, for example, the curved lines of the letter *O* contrast with the angled straight lines of the letter *X*. Since the letter *O* and the letter *X* are equally salient, some other means must be invoked to select one or the other for further processing. The other

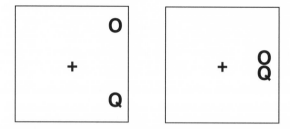

*Figure 5.3.*   High similarity of target and distractor would be expected to affect resolution processing *(right display)* but not orienting *(left display)*.

means is assumed to be top-down attentional selection of attributes and/or location generated from temporary object files (Kahneman, Treisman, and Gibbs, 1992) in working memory. The attentional selection operations arising from working memory are assumed to involve thalamic circuitry to sharpen and enhance their potency as they are projected to posterior cortical areas, where this attention to attributes and location is expressed.

Orienting and selection operations have been contrasted here by means of a task that involves a "long-distance" shift, at least between the fixation object and the target object. Many other tasks, common to daily life, involve "short-distance" shifts of less than a degree or so, such as during the examination of parts of an object. Still other tasks may involve no shift of attention, such as detecting the change in illumination of the moon behind a cloud or listening to the notes played by a particular instrument within an orchestral piece.

One might characterize the computational goal of orienting as noticing a new source of stimulation. After an information source has been noticed (through the orienting system), additional computations of a different kind are necessary in order to "realize" something about that new source of information, for example, its identity, category, color, or its velocity of movement. But noticing an information source is not sufficient by itself to allow realization processes to proceed. Processes that lead to noticing were distinguished from processes that realize what is being noticed in Chapter 3, where it was conjectured that repeated abrupt onsets of high-intensity stimuli can maintain the system in a state of noticing while blocking attention to realizing attributes of the signal or of other aspects of the environment. The computations that follow noticing are assumed to involve an enhancement of target object information, not only when it is surrounded by similar distractors (for example, the target O in the GOQ display) but also when it occurs without distractors nearby (for example, the brightening of a square in the orienting task developed by Posner). Thus, even if the target of the orienting process does not require further selection operations to resolve information coming from the target, additional enhancement of activity may be needed so that the realization process may proceed effectively. For example, when a moving object in the sky catches your eye (that is, when you notice it), you may not realize it is a bird if you are attending intensely to an ongoing conversation. The reason for this is that when attention is engaged in realizing what is being said, top-down control of attentional enhancements is assumed to be directed to these realization processes and not to visual realization processes. Hence, the bird in the sky may induce orienting or noticing when it "catches the eye," but subsequent identification and other realizations may not automatically follow.

These enhancements of an object when presented alone or in a cluttered field are assumed to be produced through thalamic circuits from prefrontal (top-down) influences. When targets are near similar distractors in the immediate surround, then additional computations would normally be required to enhance the target without enhancing the surround, thereby preventing the information from the distractors from reaching the modules that process realizations of the stimulus.

## The Thalamus

In addition to the superior colliculus, another subcortical structure that has been proposed as a mechanism of attention is the thalamus. Not only is the thalamus directly connected with the cortical areas in which attention is presumed to be expressed, but cortical columns themselves are included in the basic or "standard" thalamic circuit, hence the name *thalamocortical circuit*. The superior colliculus, on the other hand, projects its outputs indirectly to cortical columns via synapses in the pulvinar. There may exist circuits that involve SC cells and cortical columns in a loop (as is the case for the thalamus), but owing to the relatively complex connectivity within the superior colliculus, such loops have not yet been clearly delineated. The cortico-thalamo-cortical loop has a relatively simple structure, and researchers now understand a good deal about how it functions in sleep, a topic which will be briefly reviewed here. Greater detail will be devoted to research on how attentional enhancement and selection may be produced by the cortico-thalamo-cortical loop during waking.

Many researchers over the past decades have suggested that thalamic circuitry is involved in the process of selective attention (Brunia, 1993; Chalupa, 1977; Crick, 1984; LaBerge and Brown, 1989; Scheibel, 1981; Sherman and Koch, 1986; Singer, 1977; Yingling and Skinner, 1977; for an opposing view, see Mumford,

1991). It has also been specifically proposed that thalamic circuits are involved in producing enhancements of processing in the cortical areas to which they project (Crick, 1984; LaBerge, 1990; Ojemann, 1983). This section attempts to gather together experimental evidence and theoretical considerations that support both the selective role of the thalamocortical circuit and the general enhancement role of the thalamocortical circuit, whereby cortical enhancements occur even when an object (or idea) is presented in the absence of neighboring distractors.

Attentional mechanisms of the thalamus are approached here from a standpoint that takes into consideration the expression of attention in the cortical pathways to which the thalamocortical cells directly project. For example, the cortical pathways within the posterior parietal areas express attention to spatial location, those within the V1-to-IT pathways express attention to object attributes, those within certain anterior cortical areas express attention to plans and actions, and all of these attentional expressions may take place concurrently. The expression of attention in cortical pathways is assumed to take the form of a difference in the information flow between cells representing a target area and its surround. In Chapter 1 this difference was described as being realized in three ways: as an enhancement at the target area, an attenuation at the surrounding area, or both. This target/surround difference is assumed to be produced from some source outside the V1-to-IT pathway that contains the spatial information indexing the target site. Chapter 4 described several brain areas whose neurons respond preferentially to the spatial location of an object in the visual field. The PPC apparently projects directly to areas V4 and TEO, while the SC projection is indirect (through the thalamus) and FEF projection appears to be indirect (through the PPC). The PPC also projects to the same areas of the thalamus that project to the V1-to-IT pathway, which means that the thalamus is in a position to influence the projections of all three of these spatial areas to the V1-to-IT pathway.

Attentional selectivity is assumed in this book to be the modulation of neural activity flowing through the thalamus onto a cortical pathway so that a relatively small difference in the target/surround information flow at the input to the thalamus is considerably augmented as it leaves the thalamus. In other words, the algorithm of selection is dominated by the enhancement of information flow at the target site. The same circuit properties that underlie selectivity also underlie the separate assumption of general attentional enhancement of the target when no distractors are present. These two assumptions, enhancement and selectivity, will be supported by drawing on findings from studies that span much of the range of cognitive neuroscience, from behavioral experiments to PET scans to simulations of neural-network operations. These topics will be better understood after a brief review of the relevant anatomical features of the thalamus.

*Structure of the thalamus.* The thalamus of each hemisphere is a part of the diencephalon, an area in the center of the brain connecting the forebrain and the midbrain, and probably received its name, from the Greek for "inner chamber," because of its location. In the human it measures about 30 mm rostral-caudally and about 16 mm both medial-laterally and dorsal-ventrally, about the size of the end joint of a typical little finger of the hand (see Figure 5.4). Strictly speaking, the thalamus is divided into three sectors (Jones, 1985; Steriade et al., 1991): the dorsal thalamus (in which all nuclei send fibers to and receive fibers from the cortex), the epithalamus (in which nuclei neither send fibers to nor receive fibers from the cortex), and the ventral thalamus (in which nuclei receive fibers from but do not send fibers to the cortex). The dorsal thalamus and epithalamus (which contains the reticular nucleus) have been implicated in attentional processing, and henceforth all references to "the thalamus" pertain to these two structures.

Figure 5.5 shows graphically the principal nuclei of the thalamus. Most familiar to the typical reader are the lateral geniculate nucleus (LGN) and the medial geniculate nucleus (MGN), which contain cells that "relay" information from the eye and ear, respectively, to the cortex. The ventral posterior lateral nucleus (VPL) relays somatosensory information to the cortex, and the ventroposterior medial nucleus (VPM) contains gustatory relays. Taken singly or as a group, these "sensory" thalamic nuclei clearly constitute a minor part of the thalamic volume. The vast majority of the thalamic nuclei project (relay) activity between brain areas, and for this reason they have sometimes been termed "association" nuclei. The largest nucleus in the human is the pulvinar, a Greek term meaning "pillow," that gives the "inner chamber" meaning of *thalamus* the connota-

# LATERAL

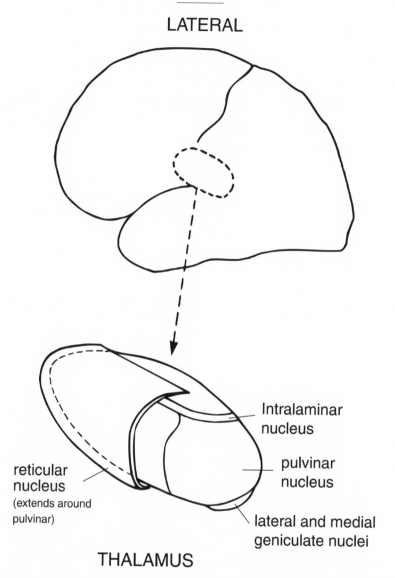

Intralaminar
nucleus

pulvinar
nucleus

reticular
nucleus
(extends around
pulvinar)

lateral and medial
geniculate nuclei

# THALAMUS

*Figure 5.4.*   The thalamus of one hemisphere of the brain. The thin layer of
neurons partially surrounding the thalamus is the reticular nucleus.

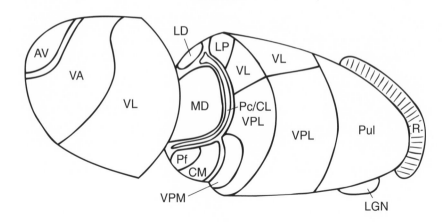

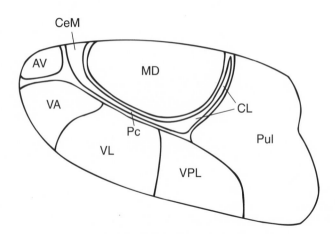

*Figure 5.5.* The main nuclei of the human thalamus *(above)* and a horizontal cross section *(below)*. Of particular interest to attentional processing discussed in this book are the pulvinar nucleus (Pul), which is directly connected mainly to areas of the posterior cortex, and the mediodorsal nucleus (MD), which is directly connected mainly to the prefrontal cortex, and the reticular nucleus (R), which extends around most of the thalamus. (Drawings are based on cross sections given by Hirai and Jones, 1989.)

tion of "bridal chamber" (Jones, 1985). The pulvinar volume is approximately two-fifths of the thalamic volume, and it has connections with virtually all of the areas of the posterior cortex and many of the areas of the anterior cortex.

The proportional size of the pulvinar is smaller in the monkey than in the human, and in the cat it is so small that it is often omitted from brain atlases. In the cat the homologous structure is called the lateral posterior-pulvinar complex. The rat thalamus contains the lateral posterior nucleus, and it is possible that this nucleus performs functions similar to the pulvinar, although at a less sophisticated level. The pulvinar evolved along with the progressive enlargement of the association areas of the posterior cortex, and the increased volume of the mediodorsal nucleus apparently follows evolutionary development of the anterior cortex. Owing to the extensive research on the role of the pulvinar in attention, it is useful to designate the four divisions of the pulvinar: inferior (PulI), lateral (PulL), medial (PulM), and anterior (PulA). The second largest thalamic nucleus in humans is the mediodorsal nucleus (MD), which serves the prefrontal areas.

Dorsal thalamic nuclei are partially surrounded (particularly on the rostral and lateral aspects) by a thin sheet of neurons called the reticular nucleus. Axon fibers that reciprocally connect the dorsal thalamus and the cortex traverse the reticular nucleus, and en route they give off collaterals that excite reticular nucleus cells. Reticular nucleus cells project to cells of the dorsal thalamus, and to each other, both axo-dendritically and dendro-dendritically (see Figure 5.6). They are known to secrete GABA at these synapses, and therefore their discharges act in an inhibitory manner on each other and on other types of cells.

*Connections of the thalamus with other brain areas.* It is well known that all sensory inputs to the neocortex are relayed through the thalamus (olfactory sensory neurons project directly to the paleocortex, or "old" cortex). In contrast to the impression given by many introductory textbooks, however, only a small fraction of the signals relayed through the thalamus arise from sensory neurons. The vast majority of signals traversing the thalamus arise from the cortex itself and from a variety of subcortical areas, notably the superior colliculus and the basal ganglia. Virtually every cortical area sends signals to and receives signals from a thalamic nucleus, and there is consider-

# THALAMOCORTICAL CIRCUIT

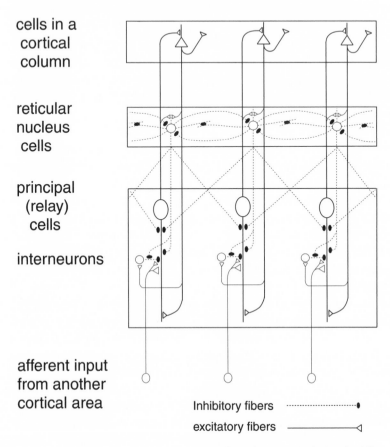

cells in a
cortical
column

reticular
nucleus
cells

principal
(relay)
cells

interneurons

afferent input
from another
cortical area

Inhibitory fibers

excitatory fibers

*Figure 5.6.* Three columns of the "standard" thalamocortical circuit. Afferent inputs (e.g., from the PPC) synapse on dendrites of principal (relay) cells in glomeruli that contain inhibitory interneuron synapses. Principal cells send their fibers mainly to layer III of a typical cortical area (e.g., V4), and returning fibers arise mainly from cortical layer VI. (Based on Jones, 1985, 1988.)

able anatomical precision in the mappings between cortex and thalamus (Jones, 1985). In general, thalamic nuclei concerned with informative signals of the sensory, motor, and association type (in contrast to the modulatory signals of the intralaminar projections to layers I and VI) send their fibers mainly to midlayers III and IV of the cortex, with minor projections to the remaining layers (Jones, 1988; Steriade et al., 1991).

Thalamocortical axons terminate on both excitatory (e.g., pyramidal and spiny stellate) and inhibitory cortical cells and are in a position to determine receptive-field properties of cortical cells (Jones, 1988). It is known that the typical direction of information flow within a cortical column begins at the middle layers and proceeds upward to layer II, then to layer V, and then to layer VI (Gilbert, 1983). Typically (but not always), outputs from layer II are sent to other cortical areas, outputs from layer V are sent to subcortical structures such as the SC and basal ganglia, and layer VI outputs return to the thalamic region from which the inputs to layers III and IV originally arose.

Of particular importance to attentional processing in the posterior cortex are the connections between the pulvinar nucleus and the areas of the occipital, temporal, and parietal lobes. Occipital areas (including areas V1–V5) project to both lateral and inferior pulvinar (Allman et al., 1972; Burton and Jones, 1976; Bender, 1981; Benevento and Rezak, 1975; Lund et al., 1981; Dick et al., 1991; Kaske et al., 1991, Ungerleider et al., 1983). Auditory areas adjacent to the primary auditory area, A1, are known to project to pulvinar nuclei (including lateral and medial pulvinar nuclei) and to the mediodorsal nucleus (Pandya et al., 1994). In the posterior parietal area, LIP cells project mainly to the lateral pulvinar, cells in area 7a project mainly to the medial pulvinar (Asanuma et al., 1985; Schamahmann and Pandya, 1990; Weber and Yin, 1984) and cells in area 7b project to the anterior pulvinar (Asanuma et al., 1985).

The lateral and inferior pulvinar contain retinotopically organized maps (Bender, 1981; Gattass et al., 1978; Petersen et al., 1985). Cortical zones of pulvinar projections show considerable overlap, compared with the striate projection zones of the LGN, a pattern that is consistent with findings of an increase in receptive-field size as one moves deeper into the visual pathways of the occipito-temporal region and occipito-parietal region. The shapes of the projection fields of striate and extrastriate neurons within the pulvinar are not straight or cone-like, as in the LGN, but rather flattened and bent in several directions, resembling discs (Dick et al., 1991). Andersen also noted disc-like shapes of projective fields of PPC neurons in the medial pulvinar (Andersen, 1987).

The medial pulvinar connects not only with the PPC but also with various areas of the prefrontal cortex (Asanuma et al., 1985; Andersen, 1987; Goldman-Rakic and Porrino, 1985), in particular the dorsolateral area (area 46), which is also implicated strongly in spatial processing (Goldman-Rakic, 1987). But parts of the medial pulvinar are also connected with areas of the temporal lobe (Baizer et al., 1991; Dick et al., 1991) and ventrolateral prefrontal areas (Goldman-Rakic and Porrino, 1985) that are implicated in the processing of objects and attributes. Thus, the medial pulvinar is in a position to mediate the flow of both spatial and attribute information between the anterior and posterior cortices, and the anterior-to-posterior flow is assumed here to constitute the top-down, voluntary control of attribute and spatial attentional processes of both selection and preparation.

The second largest human thalamic nucleus is the mediodorsal nucleus, which lies about halfway along the rostral-caudal dimension of the thalamic mass and near the brain midline (see Figure 5.5). This nucleus has extensive interconnections with the prefrontal areas, including the dorsolateral area (area 46) and the ventrolateral areas (Goldman-Rakic and Porrino, 1985). Unlike the thalamocortical connections in the posterior cortex, however, the thalamocortical connections in the frontal cortex (including the prefrontal cortical areas that connect with the mediodorsal nucleus), are under the tonic control of inhibitory input from the basal ganglia. The fact that the brain is designed to mediate basal ganglia control over the anterior cortex through the thalamus instead of directly to cortical layers provides additional support for the claim that processing in a given cortical area is intimately related to activity in the particular thalamic nucleus with which it has reciprocal and close connections.

*Attentional effects found in the pulvinar nucleus of the thalamus.* Several physiological measures have suggested that the pulvinar is responsive to tasks that involve attentional operations. Single-cell recordings in monkey pulvinar cells showed an enhancement of cell firings to a visual stimulus when that stimulus was a target of an impending eye movement or when the animal attended to it without a subsequent eye movement (Peterson et al., 1985). Using a spatial-orienting task, Petersen et al. (1987) injected muscimol, a GABA agonist (it potentiates inhibitory effects at synapses), into the dorsal region of the pulvinar and found that the ability to shift attention to the contralateral visual field was impaired, while injections of the GABA antagonist bicucilline into the same area facilitated the shift of attention to the contralateral visual field.

Rafal and Posner (1987) used the same spatial-orienting task with human patients with lesions of the posterior thalamus on one side, and found that the patients were slower to respond to visual stimuli (cued as well as uncued) in the field contralateral to the lesion though they showed no signs of contralateral neglect. The results were interpreted by the authors as indicating an impairment in engaging attention at a new location, not an impairment in disengaging attention, which is characteristic of the neglect syndrome typically produced by lesions in the PPC (e.g., Posner et al., 1984). Lesions in monkey pulvinar have also been shown to produce impairments in the attentional scanning of a visual display (Underleider and Christiansen, 1979).

Elevated pulvinar activity during attention tasks has been observed in normal humans by positron emission tomography (PET). LaBerge and Buchsbaum (1990) used a task that intensified the attentional operation of selecting a target shape when that shape was closely surrounded on all sides by similar shapes. The target was the letter O, and the surrounding distracting letters were G and Q (see Figure 5.7). On half the trials the O was replaced with a C or Ø, and the subject responded with a button press when an O appeared. As a control condition, the other displays contained a single O (without distractors) that was the same size (and had the same number of pixels) as the display with eight distracting symbols. On a given trial, only one display appeared, either the eight-distractor display or the no-distractor display. Seven subjects participated in two sessions, separated by approximately a week. In one session the 8-distractor display appeared to the left of a center fixation point, and the no-distractor display appeared to the right of center; in the other session the sides of the two display types were reversed. As the subjects began the task, they were injected with radioactive glucose (FDG) and continued the task, with short rest periods inserted, for approximately thirty-two minutes. Fifteen minutes after performing the task, they underwent a PET scan that consisted of six overlapping horizontal scans in the region of the thalamus.

The output from each subject's PET scan in terms of brain coordinates was mapped on that subject's MRI so that brain regions of

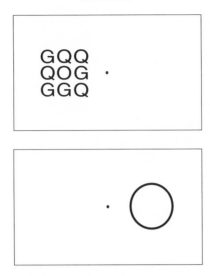

*Figure 5.7.* Two types of stimulus displays presented in a PET study (LaBerge and Buchsbaum, 1990) designed to determine pulvinar involvement in visual selective attention. Subjects fixated on the central dot and pressed a button when an O appeared at the center of the nine-item group (which was always presented to the left) or when an O appeared to the right. Half the time the O was replaced by a C or a Ø, to which subjects were not to respond.

interest could be evaluated as accurately as existing technology would allow. The results showed that, on average, the glucose uptake in the pulvinar contralateral to the side of the 8-distractor display was significantly greater than in the pulvinar contralateral to the side receiving the no-distractor display. A comparison of the corresponding relationships for the second largest thalamic nucleus, the medio-dorsal nucleus, showed no significant difference, indicating that the effect found for the pulvinar was not a general thalamic effect. Finally, a comparison of the corresponding relationships in area V1 showed, if anything, a trend toward greater glucose uptake on the side contralateral to the no-distractor display (i.e., a trend opposite to that found for the pulvinar). Therefore, the obtained pulvinar difference could not have arisen from a direct projection of an activity difference in area V1. Taken together, these results suggest that circuits within the pulvinar may operate to select the location of a visual shape when other shapes are positioned close by.

Recently a closely related PET experiment was performed with normal humans in the Brain Imaging Center at the University of Texas, San Antonio (Liotti, Fox, and LaBerge, 1994). By using labeled water instead of glucose, the researchers could test different conditions in the same session, because labeled water washes out rapidly. The same attention-demanding stimulus was used here as was used in the related PET experiment (LaBerge and Buchsbaum, 1990). The least demanding stimulus was the letter O surrounded by slanted lines, and subjects pressed a button only when the display contained the letter O and not when a C or a Ø appeared in its place. In a block of trials only one type of distractor appeared, either the Gs and Qs (letter distractors) or the slanted lines (line distractors), and as each trial began a cue indicated which side of center the cluster of items would appear. Subjects maintained fixation at the center plus sign at all times. A third condition was a control, in which no target appeared and subjects responded randomly when the neutral cue went off.

The results showed significant increases in blood flow in the pulvinar when the PET activations of the low-attention task (line distractors) were subtracted from those of the high-attention task (letter distractors), when those of the neutral task were subtracted from the high-attention task, and when those of the neutral task were subtracted from the low-attention task. Significant changes were also observed in the mediodorsal nucleus, and also in areas corresponding to the V1-to-V4 pathway, the posterior parietal lobe, the dorsolateral prefrontal area, and the anterior cingulate area. Most of these areas have been strongly implicated in single-cell recordings and other PET studies of visual attention, as described in Chapter 4.

Of particular interest in the results of the Liotti, Fox, and LaBerge (1994) study is the finding that the thalamic nuclei showed very high changes in blood flow between the high- and low-attention conditions, and that the absolute level of blood flow in the thalamic nuclei was very high. The task that was apparently responsible for the pulvinar effects also produced response-time effects that were measured before and during the PET sessions. An independent measure of the attentional effectiveness of the letter distractors and line distractors was obtained by occasionally inserting a single O (versus C or Ø) as a probe, to which the subjects responded with a button press. The probe occurred on the side opposite to the side that was cued, following a similar procedure developed in an earlier study (LaBerge, 1973). When the probe occurred during the letter-distractor task, the mean response time to the probe was at least 70 msec longer than when it occurred during the line-distractor task for all subjects. This difference in probe response time indicates that attention was much more strongly committed to displays having letter flankers than to displays having line flankers. Apparently, the time to disengage attention from the expected location was increased by the intensity of preparatory attention given to that location. Thus, the response-time measure provides an independent indicator of the attentional demands of these tasks.

Another PET study (Corbetta et al., 1991) revealed thalamic activity while human subjects performed a task that focused attention on one of three attributes of objects. As one part of a matching-to-sample task, the subjects were cued to discriminate an array of shapes with respect to their shape, size, color, or velocity of movement. Each pair of possible sizes, colors, and movement velocities were quite similar to each other, so that considerable concentration of preparatory attention on the target attribute would seem to have been exerted by the subjects during the period of time following the onset of the sample stimulus. However, it would seem that attention would be directed not so much to selective areas occupied by the individual objects as to the color, sizes, and velocities of the group of objects. Furthermore, attention could be directed to object locations and attributes at the center of the visual field, which may require less attentional intensity than when attention must be directed toward the periphery (see Figures 3.2 and 3.3) Nevertheless, the close similarities of colors, sizes, and velocities used in this study would suggest some pulvinar (and mediodorsal, serving top-down operations) activity involved in the attentional expressions of these attributes in extrastriate pathways.

The results of this study showed increased blood flow for the velocity, color, and shape discriminations in the right thalamus/superior colliculus region, which is presumed to include the pulvinar. Areas of the superior colliculus and thalamus that were separated in the analysis showed significant blood flow only for shape. Strong effects were found in several extrastriate areas presumed to specialize in the processing of color, shape, and velocity.

As another part of the experiment that used a divided-attention condition, the subjects were not cued prior to the discrimination display and instead of attending primarily to one of the three stimulus attributes they may have been attending to something else— for example, the upcoming actions of scanning the discrimination procedure of each attribute. The PET results from this condition showed no increase in thalamic or SC blood flow, but rather showed an increase in blood flow in the right anterior cingulate area and right dorsolateral prefrontal area. Apparently the thalamus showed detectable involvement only when attention was focused on a visual stimulus attribute. One explanation for the lack of measured pulvinar effects in the divided-attention condition is that the duration of time that attention is directed to a particular attribute (color, shape, or velocity) on a trial is likely to be confined to part of the time that the display is shown, when the subject tested each attribute against the sample display. In the focused-attention condition, the attribute to be examined can be attended to beginning with the cue onset and can continue when the display appears. Thus, not only can the attentional networks involved in processing a given attribute be active for a longer time in the focused-attention condition than they are in the divided-attention condition, but the longer duration may allow attention to be built up to a higher intensity.

Not all recent neurobiological evidence favors the conclusion that the pulvinar is involved in visual attention of the selective and preparatory type. A drug-injection study questions whether the pulvinar is necessary for the selective processing of a target object when another object is presented at the same time, either in the same or in the opposite hemifield (Desimone et al., 1991). In this study, the lateral pulvinar on one side was deactivated by injections of muscimol, and a colored bar and a distracting bar (positioned about 2 degrees apart) were presented in the visual field contralateral to the affected pulvinar. Pulvinar deactivation had no effect on performance in comparison with the results when no distractor was present. When the SC was unilaterally deactivated with injections of mucimol, however, performance was impaired when a distractor appeared in either the same or opposite visual field (all in the absence of eye movements). This finding is consistent with the observation that lesions in the superior colliculus increase the time needed to shift attention (Kertzman and Robinson, 1988; Posner and Cohen, 1984).

One may attempt to formulate a consistent account of these apparent discrepancies by examining carefully the characteristics of the visual displays used in the behavioral tasks. If one assumes that attention to location makes use of eye-movement information, as has been assumed in the present discussion of computations in area LIP and the SC, then disruptions in the eye-movement computations in the superior colliculus would be expected to feed back to area LIP and cause disruptions there as well. The Desimone et al. (1991) experiment suggests that with their displays, in which distractors are positioned at least 2 degrees from the target and discriminations are based on a single feature (e.g., color), pulvinar processing may be bypassed, perhaps by the direct route that exists between area LIP (which specializes in oculomotor information) and V4 (Asanuma et al., 1985; Fries, 1984; Lynch et al., 1985). In tasks where the spacing between the target and distractor is potentially large enough to evoke different eye movements, the mechanism that produces the expression of attention in V4 would seem to be closely associated with oculomotor information used by the superior colliculus and shared with area LIP. Additional research is needed to determine clearly the relationships between target/distractor distances and the involvement of selection mechanisms attributed to oculomotor information (associated with the SC) and close spatial and attribute discriminations (conjectured here to be associated with the thalamus).

## Thalamic Circuitry

While neurobiological experimentation continues to accumulate data that may lead to clarifications of attentional mechanisms, one can also put a candidate mechanism to the test by determining whether its circuitry may instantiate a selective attention algorithm. In other words, much may be learned by discovering whether a circuit has the potential for converting input firing patterns into output firing patterns that are appropriate for expressing attention in cortical pathways. But before an effective simulation can be carried out, one must first have reasonably clear knowledge of the circuit network on which the simulation is based. Current knowledge of the circuitry of the superior colliculus is rather sparse, although much is known of its connectivity with other brain structures and the firing behavior of its cells in a variety of visual tasks (e.g., Sparks and Mays,

1980; Wurtz and Goldberg, 1972a; Mohler and Wurtz, 1976). In contrast, much is known not only of the structure of the thalamic circuitry but also of the firing characteristics of the component neural units. I will attempt now to describe briefly the "standard" thalamo-cortical circuit structure that characterizes virtually all thalamic nuclei (Jones, 1985), and then I will describe a study that simulated the operation of this circuit in the pulvinar in the context of a spatial attention task.

*The structure of the standard thalamic circuit.* Almost all of the knowledge we have of the circuitry of the thalamus is based on neurobiological studies of the monkey, cat, and rat. The circuitry of the typical thalamic nucleus is apparently quite similar across these species, and comparisons of histochemical staining in the monkey and human thalamus show identical patterns (Jones, 1985; Steriade et al., 1990). It seems reasonably safe, therefore, to generalize many findings in rat, cat, and monkey thalami to the human thalamus.

The relative simplicity of the thalamic circuitry is a consequence of the column-like structure of cells and synapses that intervene between the input and output of the thalamus and of the fact that these columns are repeated from nucleus to nucleus across the thalamus (Jones, 1985; Steriade et al., 1990). A schematic diagram of three adjacent typical columns is shown in Figure 5.6, revealing the major components of the thalamic circuit. Depending upon the particular nucleus, the thalamic inputs (shown at the bottom of the diagram) arise from sensory afferents (e.g., from the LGN and VPL), from SC afferents, and from cortical afferents (e.g., the input to the pulvinar arises mainly from posterior cortical areas, and the input to the mediodorsal nucleus arises mainly from the many prefrontal cortical areas). The axons of the afferent inputs terminate on the two types of cells that constitute the dorsal thalamus: the principal (relay) cells and the interneurons. The output of the thalamic circuit is produced by axons of the principal cells, which project directly to cells in a column of a particular cortical area. Axons of principal cells do not contact each other, but as they pierce the reticular nucleus en route to the cortex they send off collateral axons that terminate on proximal dendrites of reticular nucleus cells (Steriade et al., 1990). It could be said that the thalamic column "serves" the cortical area to which the principal cells project. Thus, the principal cells of a column provide channels of information flow through the thalamus to the cortex.

*Output connections of the thalamus to cortical areas.* The principal thalamic cell projects to more than one layer of cortex. A traditional
belief holds that middle layer IV is the main target of thalamocortical
projections, but this may be the case only for area V1, the first auditory area, and somatosensory areas 3a and 3b (at least in monkeys);
for most cortical areas, the main target of thalamocortical projections is layer III (Steriade et al., 1990). When a thalamocortical axon
enters a cortical layer, it may synapse with all neural elements in this
layer, including dendritic shafts and spines of pyramidal cells and
spiny and aspiny stellate cells (Steriade et al., 1989). As already indicated here, the typical route of activity flow within a cortical column
proceeds from middle layers III and IV to layer II, then to layer V,
and then to layer VI (Gilbert, 1983), where cells project not only
back to the thalamus but also back to layer IV (Steriade et al., 1990;
Ferster and Linstrom, 1983). Thus the loop extending from the thalamus to the cortex and back to the thalamus contains within it
another loop (at least) that extends from the middle layer to the
deeper layers and back to the middle layer. Since the connecting
cells of these loops are excitatory, the loops would seem to be capable of sustaining a signal arriving by way of a thalamic afferent fiber.

All cortical layers can project via pyramidal axons to sites outside
the cortical column: layers II and III to other cortical areas, layer V
to subcortical regions (e.g., the SC, basal ganglia), and layer VI to
the thalamus. Although thalamocortical projections to any of the six
cortical layers can eventually return activity to the thalamus by the
routes to layer VI, layer VI provides the most direct feedback loop
to the thalamus via thalamocortical cells that terminate on layer VI
cells (Steriade et al., 1990).

The principal cell axons apparently project with some precision
to a cortical area. In a typical cortical area of extrastriate cortex,
branches of individual axons appear to be mainly confined to regions covering approximately 2–2.5mm (Kaska et al., 1991), which
is only slightly larger than the 1–2 mm projection fields reported
for the primary visual, auditory, and somatosensory areas (Ferster
and Levay, 1978; Gilbert and Wiesel, 1983; Redies et al., 1989). It
would appear therefore that the topographical precision provided
by the separation of thalamic columns is offset somewhat by the distribution of principal cell terminals in the cortex. However, the extent of this distribution seems quite narrow relative to the size of a
typical architectonic cortical area.

The projection from cortex back to the thalamus occurs mostly by way of layer VI cells (the layer V projections to thalamus will be discussed later, under the section on triangular circuits). The excitatory axons of layer VI cells, which terminate on distal dendrites of the thalamic principal cells, provide approximately half of the principal cell synapses, at least in the LGN (Sherman and Koch, 1990). The functional significance of this connectivity pattern may lie in the ability of distal dendrite activity to shift the soma membrane closer to threshold so that afferent impulses arriving nearer the soma are potentiated (McCormick, 1992). This "short-term potentiation-like" effect within the thalamus could be viewed as modulating the incoming afferents to the thalamus.

*Inputs to the thalamus in triangular circuits linking cortical areas to each other.* We have examined in some detail the structure and functions (both known and conjectured) of the basic thalamic circuit, particularly for the pulvinar nucleus, and we have noted that the circuit appears to be structured to influence cortical activity in a sufficiently strong way to produce an expression of selective attention. We turn now to the evidence that indicates the cortical sources of the main afferent inputs to the pulvinar circuits within the thalamus. The afferent inputs to the pulvinar circuit from various cortical areas (such as the PPC) are regarded here as the controls on the circuit mechanism that could produce the expression of attention in a localized region of another cortical area (e.g., V4).

The external fibers that project to a thalamic nucleus from cortical areas other than the one to which the nucleus sends its fibers synapse on structurally identifiable dendritic sites on principal cells close to the soma. They are typically in contact with inhibitory interneurons within glomeruli (see Figure 5.6). These projections apparently are not reciprocated to the cortical area of origin, as projections that terminate on distal dendrites are. Axons that terminate at the proximal synaptic sites are generally larger in diameter, and their boutons contain larger vesicles than those found in the axons that terminate at the distal synaptic sites (Sherman and Koch, 1990), so that they are usually distinguishable from axons that terminate on distal dendrites.

Cortical projections to pulvinar cells originate from two layers in the striate cortex: large pyramidal cells in the upper half of layer V and small and medium-size pyramidal cells in layer VI (Conley and Raskowsky, 1990); the projections from the striate cortex to the LGN originate only from layer VI. The projection from V1 to the inferior pulvinar is topographic (Benevento and Fallon, 1975). Some cells there have relatively small receptive fields, and most cells there appear to represent the foveal region rather than the periphery (Dick et al., 1991). It is presumed that the layer VI cells that send their axons to the LGN terminate on distal dendrites of principal (relay) LGN cells, and the layer V and deep VI cells that send their fibers to the pulvinar presumably terminate on proximal dendrites of principal cells. The pulvinar cells that receive projections from area V1 presumably serve an extrastriate area upstream by reciprocal connections to that area.

Other evidence for afferent cortical projections to a thalamic nucleus was provided by injecting different types of label into areas TEO and TE in the inferotemporal region (Webster et al., 1991). Labeled cells were segregated in the pulvinar, but about 20 percent of terminal label from one cortical area overlapped pulvinar cells labeled from the other cortical area (Ungerleider, pers. comm.). These findings indicate that cells in TEO project not only to pulvinar cells in areas where cells return those projections (in the manner of the basic corticothalamic loop) but also to pulvinar areas where cells receive and return projections to cells in TE (in the usual manner). Thus it appears that TEO projects to TE over two routes: the indirect (bisynaptic) route, through pulvinar cells, and the direct (monosynaptic) route, constituting a triangular circuit.

Given the suggestive evidence from areas V1 and IT for corticothalamic fibers that project to areas outside their own cortico-thalamo-cortical loops, it seems plausible to expect the same pattern to exist among pairs of other cortical areas in the V1-to-IT pathway, as diagrammed in Figure 5.8. Many labeling studies of cells in visual cortical areas show axon labeling in more than one pulvinar area (e.g., Dick et al., 1991; Kaske et al., 1991). There are also indications of triangular circuits connecting parietal and prefrontal areas involving the medial pulvinar and mediodorsal nuclei of the thalamus. Injections of DLPFC cells label terminals in the medial pulvinar and in the mediodorsal nucleus (Selemon and Goldman-Rakic, 1988), and injections of 7a cells (in the PPC) label medial pulvinar cells (Asanuma et al., 1985) and mediodorsal cells (Selemon and Goldman-Rakic, 1988). Clear evidence apparently does not yet exist that DLPFC terminals in the medial pulvinar overlap cells there that

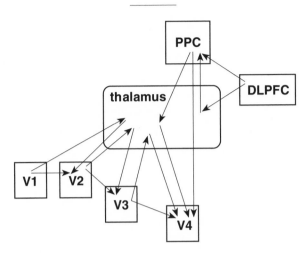

*Figure 5.8.* Conjectured triangular circuits connecting pairs of cortical areas.

project to the PPC, or that PPC terminals in the mediodorsal nucleus overlap cells there that project to the DLPFC (but see Asanuma et al., 1985; Selemon and Goldman-Rakic, 1988). Schwartz et al. (1991) have examined synapses within the region of the mediodorsal nucleus that receives projections from the DLPFC, however, and found two types of axon terminals, distinguished on the basis of the sizes of the vesicles that they contain. The large round vesicles appear in axons that terminate in glomeruli, and the small round vesicles appear in axons that terminate outside the glomeruli. In addition, they found evidence indicating that layer V axons contain the large vesicles and layer VI axons contain the small vesicles. This evidence for the dual mode of corticothalamic terminations in the prefrontal area is consistent with the findings of Conley and Razkowski (1990) in layers of the striate cortex.

The present view of the function of the pulvinar in spatial attentional processing puts considerable emphasis on an assumed triangular circuit between an area or areas of the PPC and some area or areas within the V1-to-IT pathway (e.g., a triangular circuit joining LIP and V4/TEO in the monkey, and its homologue in the human). The direct (backward) corticocortical projection from the PPC to V4/TEO is not expected to magnify small target/surround differences in the PPC—it might even decrease such differences, owing to the relatively large receptive fields in deep columns from which the PPC-to-V4/TEO projections originate and at which they terminate (forward projections typically originate in upper layers and terminate in middle layers, where receptive fields are relatively smaller). However, the other triangular route from PPC to V4/TEO traverses the pulvinar, a path that presumably enhances and sharpens the projection pattern, and the output is sent from the pulvinar to layer III of V4/TEO, a layer whose cells have smaller receptive fields than those of the deep layers.

It is hard to ignore the observation that layer V cells of the cortex tend to be "bursting cells" (Wang and McCormick, 1993) that are particularly suitable for driving the temporal patterning of thalamo-cortical cells serving other cortical areas. More specifically, it has been suggested that bursting cells may amplify and synchronize cortical outputs (Chagnac-Amital et al., 1990) and may serve to bind feature representations in disparate cortical areas through synchronous firing patterns (e.g., Crick and Koch, 1990; Singer, 1990). In general, there would seem to be computational advantages in combining temporal binding with spatial sharpening by routing cortico-cortical projections from bursting cells through a thalamic circuit.

Afferent inputs to the thalamus typically synapse within glomeruli located on proximal dendrites, where the afferents can strongly affect activity at the soma (Sherman and Koch, 1990; Jones, 1985). A glomerulus is a group of synapses surrounded by glia. In Figure 5.6, the afferent terminals synapse within a glomerulus on both a principal cell dendrite and on interneuron dendrites. Interneurons synapse with each other and with principal cell dendrites in inhibitory dendrodendritic profiles. The effects in inhibitory-to-inhibitory synapses are apparently much stronger (approximately twenty times as powerful) as effects in excitatory-to-excitatory synapses (Grace and Bunney, 1979). The structure of a glomerulus appears to enable incoming afferent signals to the thalamus to be closely modulated by inhibitory discharges of interneurons, which could shape incoming pulse trains.

*Interconnections between the thalamocortical "columns."* The feedforward inhibitory action of interneurons on the firing of principal thalamic cells is supplemented by the feedback inhibitory action of reticular nucleus cells on these same cells. Neurons of the reticular nucleus (RN) are embedded in a mesh or network of fibers (having the appearance of the reticulation of veins in a leaf). The axon of a typical RN cell sends a few collaterals to nearby RN cells and then projects into the dorsal thalamus, where it branches extensively (Yin et al., 1985). Studies of both monkey and cat brains show that RN axons inhibit principal cells and apparently also inhibit interneurons, which are themselves inhibitory (Montero and Singer, 1985; Steriade, Domich, and Oakson, 1986), at least in the LGN. It is estimated that the proportion of RN synapses in the LGN that are on interneurons is 8 percent (Takacs, Hamori, and Silakov, 1991); corresponding estimates for other thalamic nuclei, particularly the pulvinar, apparently have not yet been reported.

The RN cells inhibit each other not only by axodendritic synapses but also by the considerable number of dendrodendritic synapses, where the extensive dendritic processes of these cells contact each other. The mutually inhibitory manner of processing in the reticular nucleus makes it difficult to infer at a glance exactly how the reticular nucleus mediates effects between the principal cells. With the help of a neural-network model, however, one can more clearly describe the computations that this circuit structure might exert upon thalamic-column interactions, and thereby begin to understand how the entire thalamic circuit might function as a mechanism of attention. I address this issue in the next section.

Although the effect of RN circuitry on thalamic-column interactions while the individual is awake is presently unclear, much is known about the general effect of RN cell discharge on principal cell activity during sleep (Steriade et al., 1990). A brief description of thalamic-circuit functioning during the general cortical state of drowsiness and deep resting sleep may provide instructive hints about its functioning during states of attention whose expression may share some of the modulatory (versus informational) characteristics of sleep states.

*Thalamocortical activity during resting sleep.* When the brain passes from the alert waking state to the deep stage of resting sleep, neural firing does not cease but rather exhibits a profound change in its pattern of discharges. The traditional objective measure of waking and sleep states are EEG waves that reveal the summed electrical activity of many neurons. During resting sleep the EEG waves show high amplitudes and low frequencies and are relatively synchronized across the many areas of the cerebral cortex; in the alert waking state the EEG waves over the cortical areas show low amplitudes and relatively high frequencies and are de-synchronized.

The global expression of resting (non-REM) sleep in cortical pathways can be decomposed into three major component rhythms: the spindle (7–14 Hz), delta (1–4 Hz), and the slow (0.1–0.8 Hz) rhythms (Steriade et al., 1993; Steriade, McCormick, and Sejnowski, 1993). The spindle rhythm appears in the first stages of sleep as the individual becomes drowsy; the delta and slow rhythms follow, as sleep deepens, in later stages. Each rhythm is generated by a known neural mechanism, and these mechanisms interact to produce the irregular waves of the typical EEG recorded during deep resting sleep. A description of each of these mechanisms may suggest what it is that resting sleep accomplishes for the system.

(1) *The spindle rhythm.* The spindle rhythm is characterized by bursts of spikes that occur in the 7–14 Hz range and last for only 1–3 seconds, with interspike "lulls" of 5–8 seconds. Spindle-shaped electrical waves are known to be generated in the thalamus (even by animals without a cortex) by single neurons. When thalamocortical cells and reticular nucleus cells are hyperpolarized (to −60 millivolts), they respond with a rebound burst of spikes (Steriade, Llinas, and Jones, 1990), the low-threshold calcium spike. During the early stages of sleep, hyperpolarization in these thalamic cells is induced by a decrease in the acetylcholine (Ach) neurotransmitter projected from the brainstem (other transmitters that project to thalamic cells and change their excitability are norepinephrine, NE, serotonin, 5-HT, and histamine, HA). The rhythmic characteristic of spike bursts is promoted by the "pacemaker" activity of RN cells that send inhibitory bursts to the thalamocortical cells, which then respond with rebound bursts that are projected directly to the cortex (Steriade et al., 1993). Since the meshwork of RN cells spreads the rhythmic bursts to virtually all thalamocortical (principal) cells, the spindle oscillations will become synchronized across the entire cerebral cortex. Thus, the burst-firing that begins as an intrinsic cellular response in reticular nucleus and thalamocortical neurons becomes a source for simultaneous spindle activity in virtually all cortical areas by virtue of the circuit properties of the reticular nucleus.

One of the most dramatic effects of shifting thalamocortical cells from the regular-spiking mode to a burst-firing mode (where a burst contains a few spikes at 250–400 Hz) is the blocking of information transfer through the thalamus to the cortex. When the information from the external world, encoded in trains of signals arising from the sensory surface, reaches the "sensory" thalamic nuclei (e.g., the LGN for vision, the MGN for audition, and the ventroposterior lateral nucleus, VPLN, for touch), rhythmic bursts of high-frequency spikes in the thalamocortical (relay) neurons apparently destroy that information, and hence the cortex is deprived of sensory stimulation. Similarly, when the information arising from one cortical area is projected to another cortical area through the "association" nuclei of the thalamus (such as the pulvinar or mediodorsal nuclei), it is blocked by the burst-firing in the thalamocortical neurons and thus prevented from combining its effects with the information flowing across the direct corticocortical connections (see Figure 5.8).

(2) *The delta rhythm.* As sleep deepens, and the membrane excitability in thalamocortical cells is decreased further, these cells exhibit a delta oscillation in which bursts occur with a frequency of 1–4 Hz. Apparently, this rhythmic train of bursts originates within the thalamocortical cell, where two types of ionic currents flowing through the membrane interact in a cyclical manner (McCormick and Pape, 1990). These burst trains are projected directly to the cortex and to the RN cells, which spread and synchronize the delta rhythm over the thalamocortical cells serving the entire cortex.

(3) *The slow rhythm.* During the late stages of resting sleep another oscillation appears, but this rhythm originates in the cortex itself rather than in the thalamus (Steriade et al., 1993). When thalamocortical neurons reduce their firing rates from the 7–14 Hz in spindle sleep to 1–4 Hz in delta sleep, their cortical target cells consequently show decreased activity. Although cortical activity is at a very low level, the cortical neurons of all types (regular-spiking and burst-firing cells) produce an oscillation of cell activity at a frequency of about 0.3 Hz (within the range of 0.1 to 0.8 Hz). Thus, the cortex's slow rhythm is generated within a network of cells, not from the intrinsic properties of particular cell types, as is the case for spindle and delta rhythms. Like the spindle and delta rhythms, however, the slow rhythm is blocked by stimulation of the brainstem Ach nuclei that raise the excitability of cell membranes in cortical as well as thalamic cells and move the brain toward the waking state.

*Interactions of the three rhythms in deep resting sleep.* The slow rhythm is apparently projected from cortical cells to thalamic cells, where it tends to group the occasional spindle and delta bursts by resetting the times of their onsets (Steriade et al., 1993). The influence of the slow cortical rhythm on thalamic activity appears to be strong, owing to the fact that most of the afferent axons entering the thalamus arise from the cortex. Thus the irregular EEG observed from scalp electrodes during the deepest stages of resting sleep reflects the co-ordinated activity of the three types of rhythms, each of which exhibits a more stereotyped rhythm when conditions allow it to be observed in isolation.

Taken together, the three rhythms of resting sleep appear to reflect a cortical state that is not only closed to the external world but also prevented from generating organized patterns of activity within itself. It has been conjectured that conditions of slow-wave sleep might promote the consolidation of information acquired during the waking state (Steriade, McCormick, and Sejnowski, 1993).

One could summarize the foregoing descriptions of the activity of thalamocortical networks during resting sleep in terms of the expression, mechanisms, and control of sleep. In the early (drowsy) stage of resting sleep, cortical activity begins to show sequences of bursts (at 7–14 Hz) organized as spindles, which presumably act to break up normal waking patterns of activity. The mechanism that produces the spindling is the low-threshold calcium spike generated in thalamic RN and thalamocortical cells, which acts at the thalamic level to block incoming sensory information that contributes to organized cortical activity during the waking state. The most pervasive source of control on this mechanism is the level of neurotransmitters (Ach, NE, 5-HT) currently being projected to these thalamic cells from brainstem nuclei. More local sources of control are the synchronizing effects of RN cells and the volleys of spikes arising from corticothalamic fibers.

As the neurotransmitter level continues to decline, the expression of sleep takes on the delta rhythm (1–4 Hz), which also consists of cyclical bursts that disrupt organized patterns of firing in the cortex, but these bursts fire at a slower rate and in a more continuous manner than those produced by the spindle rhythm. The mechanism that generates the delta rhythm is the particular membrane property of the thalamocortical cell. The excitability of this intrinsic cellular structure is controlled by the level of neurotransmitter projected from nuclei in the brainstem and by volleys arising from RN cells and corticothalamic cells.

Finally, in deepest sleep, the slow wave ($<1$ Hz) becomes the dominant rhythm, revealing a cortical state that is apparently devoid of organized spatial or temporal activity other than the infrequent synchronous pulsing of almost all of the neurons of the entire cortex. The mechanism that generates this slow rhythm is the network interaction of virtually all types of cortical cells. The control of this type of low-level interaction is not only a low level of Ach at the synapses but also a decrease in the barrage of thalamic bursts resulting from the reduction of spindling during later stages of resting sleep.

The research that produced the foregoing conclusions and conjectures emphasized mechanisms at the cellular level during the period of time that spindling and delta rhythms were regarded as the main markers of resting sleep. When the slow wave was recently discovered, however, the mechanism controlling it was found to be at the network level rather than at the cellular level. Furthermore, the cellular mechanisms of spindling and delta rhythms are modulated by not only by returning volleys from the cortex but also by bursts from local RN cells that are involved in synchronizing rhythmic activity across the entire cortex. Thus, there appears to be a noteworthy shift toward network-level analyses of the mechanisms underlying resting sleep. The following quote (Steriade et al., 1993, p. 3298) underscores this shift in the level of analysis of natural resting sleep and suggests similar considerations of levels of analysis during waking: "The association between the three types of oscillatory activities at the single-cell and global EEG levels, as they occur during natural sleep, emphasizes the need of a preparation with preserved circuitry when analyzing the interactions between thalamic and cortical networks."

*Thalamocortical activity compared during sleep and waking.* The thalamocortical circuitry involved in resting sleep is the same circuitry assumed in this book to be involved in attentional activity during waking. When the brain shifts from resting sleep to waking, however, brainstem nuclei increase the levels of neurotransmitters such as Ach, NE, and 5-HT, which increases the excitability of thalamic neurons. The increased excitability, expressed by changes in ion conductances through the cell membranes, shifts the firing of the thalamic cells from a burst-firing mode to a tonic-firing mode. In the thalamocortical cell, the tonic-firing mode can follow the input pattern of spikes arriving from sensory surfaces or from other cortical areas, thus preserving the information in these signals as they pass through the thalamus en route to the cortical columns.

A central hypothesis of this book is that certain patterns of incoming signals can be sharpened as they pass through the thalamocortical circuit so that a small difference in firing rates between one cluster of signals and its surround entering the circuit will leave the circuit greatly magnified. The appropriate level of description and analysis of attentional operations appears to be the circuit level, even if particular cells within this circuit may have specialized firing characteristics, as was the case for thalamic cells in the generation of sleep rhythms.

If it turns out that the thalamocortical network is a mechanism that generates the expression of attention in cortical pathways, much as it generates the expression of resting sleep in cortical pathways (at least in its early stages), then much of what is learned about circuit anatomy, temporal firing patterns, and intrinsic properties of neurons within this circuit during sleep should be helpful in understanding how the thalamocortical network operates during attention states.

One salient difference between thalamocortical operations during sleep and attentional states of waking is that attention during waking apparently operates much more locally on cortical pathways than sleep mechanisms. The diffuse spread of activity during sleep is a consequence of the meshwork structure of the reticular nucleus (in which connections between cells include dendrite-to-dendrite as well as axon-to-dendrite and axon-to-soma) combined with the lowered firing threshold produced by the withdrawal of Ach and other neurotransmitters in resting sleep. The RN cells are known to inhibit each other, and when inhibition hyperpolarizes an RN cell sufficiently, it produces a rebound burst. In this way a network of connected RN inhibitory cells can spread activity to every cell within the network, apparently without decrement in the intensity of the activity.

During waking, the cells of the reticular nucleus shift to the tonic-firing mode in which cells inhibit the firing of cells to which they are connected in the typical manner. When attentional effects raise the activity in a cluster of thalamocortical cells, they activate local

RN cells, which in turn inhibit neighboring RN cells and thereby restrict the spread of activity across the reticular nucleus. Thus, the tonic-firing mode of RN cells during waking constrains the effects of selective attention to local columns of a given cortical area, while the burst-firing mode of RN cells during resting sleep diffuses burst activity throughout the reticular nucleus and therefore throughout virtually all thalamocortical cells of the brain.

Although the expression of attention is defined within a local area of cortex, several different areas of cortex may express attention simultaneously in a coordinated way. For example, when one is attending to the middle letter of a three-letter word, presumably the expression of attention at the center location takes place not only in pathways of the posterior parietal cortex (PPC) but also in pathways of area V4, where shape information arising from the center location is enhanced relative to the distractor information. The coordination of the attentional expressions in these different cortical areas is assumed here to be simultaneous amplification patterns. Synchronized patterns of firing, perhaps at or around 40 Hz (Eckhorn et al., 1988; Gray and Singer, 1989), suggest another way that separate expressions of attention may be coordinated.

In summary, the major "computational" goal of resting sleep appears to be the blocking of sensory information from reaching the cortex and the prevention of information processing within and between cortical areas. The brain achieves this goal first by breaking up ongoing cortical processing by barrages of spikes in the form of burst spindles, generated in thalamic cells, that are synchronized across the entire cortex by means of the reticular nucleus of the thalamus. The spindles of bursting also effectively block information transfer through the thalamus from sensory surfaces and from other cortical areas. As sleep deepens, the synchronized firing of massive groups of cortical cells continues in a somewhat less intense form as the thalamic-generated delta rhythm and the cortex-generated slow rhythm become dominant.

*Thalamocortical circuits and the intensification of attention.* The previous section described how the thalamocortical circuit operates in an oscillatory burst-firing mode that generates resting sleep. For decades introductory textbooks have described a second mode that operates during waking to "relay" information arising from sensory surfaces to the cortex. It appears now that the relay function during waking also applies to the many thalamic inputs that arise from one cortical area and are sent by the thalamus to another cortical region. However, the thalamocortical circuitry, with its known lateral inhibitory and recurrent excitation components, appears to have a more sophisticated function than simply relaying information from the senses to the cortex or from one cortical area to another cortical area. A major goal of this book is to explore the hypothesis that the thalamocortical circuit operates in a third mode, which is the attentional intensification of local activity in a variety of cortical areas.

As mentioned earlier in this chapter, the function of selective attention has been ascribed to thalamic circuitry by several investigators (e.g., Chalupa, 1977; Crick, 1984; LaBerge and Brown, 1989; Scheibel, 1981; Singer, 1977; Sherman and Koch, 1986; Yingling and Skinner, 1977; for a recent computational model involving the pulvinar, see Olshausen et al., 1993). Skinner and Yingling (1977) proposed that attention operates by a gating mechanism (selection by inhibition of the non-attended object sites) within the sensory nuclei of the thalamus, and Sherman and Koch (1986) proposed a selection mechanism that operated within the LGN. Crick (1984) proposed that the "searchlight" of attention is expressed by bursting cells in the LGN relay cells while the RN cells inhibit firing in neighboring LGN cells. It is now known that the burst-firing mode is characteristic of sleep, not of waking, and that the top-down control of selective attention in early visual processing is anatomically more direct at the pulvinar nucleus than at the LGN. If the LGN were involved in visual attention, then one would expect to see attentional effects in area V1, which the LGN projects to directly. Only a few studies have found attentional effects in area V1 (e.g., Motter, 1993), however, and these effects are not as strong as those found in higher visual areas, such as V2, V3, or V4. Meanwhile, there is relatively direct evidence of pulvinar involvement in visual attention from single-cell recordings and PET scans (e.g., Corbetta et al., 1991; LaBerge and Buchsbaum, 1990; Liotti et al., 1994) but no parallel evidence of LGN involvement in visual attention.

Therefore, the nucleus of the thalamus that appears most likely to be involved in early visual attention is the pulvinar. The thalamic nucleus that is connected both to prefrontal areas assumed to be involved in the control of attention (e.g., the DLPFC, inferior convexity, the supplemental motor area, or SMA, and the anterior cingulate area) and to visual areas in the posterior cortex is the mediodorsal nucleus. Since the form of the thalamocortical circuitry is virtually the same in the pulvinar and mediodorsal nuclei, all discussions of attentional operations in the thalamocortical circuit apply to both nuclei.

*Simulations of thalamic circuit operations.* In order to determine whether or not the standard thalamic circuit could operate in the "attention mode," we carried out simulations of the operations of three network models of the basic thalamic circuit. Results of a neural-network simulation does not prove that the corresponding neural circuit in fact produces the same output. Rather it shows that the neural circuit under consideration could produce the output pattern shown by the simulation. Thus, a simulation provides something akin to an existence proof of a class of input-output relationships generated by a particular neural circuit.

Although the general anatomical structure of the pulvinar circuit and thalamic circuits in general is known, there remains some uncertainty about the connections between the reticular nucleus (RN) cells and the principal cells and interneurons in the dorsal thalamus. At present there is no clear evidence that an RN cell projects only to the principal cells that innervate it, or only to principal cells in neighboring columns, or to both. Furthermore, it is not known how strongly RN cells project to interneurons. Since these alternative projection patterns of RN cells into the dorsal thalamus appeared to span the range of plausible RN projections, we decided to model all three cases of the RN-to-principal-cell patterns, and we included the RN-to-interneuron connection in the version in which the RN cell projects only to the principal cell that innervates it. These three network models of the thalamic circuit are shown in Figure 5.9. To visualize a particular network within the larger context of brain architecture (e.g., Figure 5.8), one may invert the network so that the afferent inputs arise from cells of the PPC and the principal-cell outputs project to cells of V4.

The equations describing the simulation (LaBerge, Carter, and Brown, 1992) are of the form used in traditional connectionist models (Rumelhart and McClelland, 1986). The response of each unit was represented by a number (between 0 and a maximum of 100) interpreted as its rate of firing in spikes per second. The output of

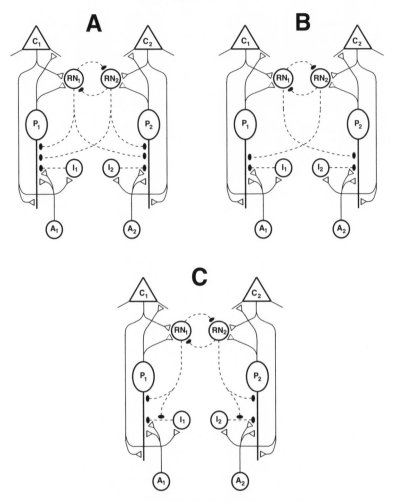

*Figure 5.9.*    Three versions of the thalamocortical circuit that vary in the way inhibitory RN axons are distributed to columns in a (dorsal) thalamic nucleus. Solid lines represent excitatory fibers and dashed lines represent inhibitory fibers. (Cell labels are shown in Figure 5.6.)

each neural unit at time $t + 1$ was computed by first summing all the inputs to that unit at time $t$ and then passing this sum through a threshold function. The threshold function had a non-linear form (the particular threshold function used here was an exponential function) that began to produce an output after the summed input value reached some minimum value. As the summed value increased, the output of the threshold function rose strongly at first and gradually leveled off as it approached the maximum value set at 100 spikes per second. Each unit was assigned a low initial value (at $t = 0$) that usually represented its base rate of firing. As time increased by 1 cycle, each unit was updated by an equation that expressed the process of summing each input and passing that sum through the threshold function. Each unit of the network had its particular input-output equation whose components depended upon which other units were connected to it. In this way the particular connective form of the thalamocortical network could be represented by the set of equations.

As is the case in typical connectionist models, the set of non-linear equations representing the combined activities of all the neural units of the network was not solvable in a closed mathematical form. That is, the firing rate of a unit at time $t$ could not be expressed as a function of $t$ and parameters such as the initial unit values (at $t = 0$). Therefore, the firing rates of the units at time $t$ were generated by numerical analysis, which involved starting each unit at $t = 0$ and generating the new value for each unit at each successive increment of 1 cycle. In this way, each unit's firing rate over time could be graphed as a trajectory (in two dimensions), and the path in $n$-space taken over time represented the trajectory of the combined units of the network. In particular, the trajectory for a unit that represented a target input could be compared over time with the corresponding unit that represented a distractor input, and these trajectories are assumed to describe the development of selective attention over time.

Even though output firing rates were available for each neural unit in the simulations, the output trajectories of main interest were the output patterns of thalamocortical units, and therefore the following observations are confined to the time course of activity in the principal thalamic units and the cortical units.

The input to the network represented the mapping of a three-letter stimulus display (e.g., TON), in which five columns corresponded to the width of each letter and three columns to the space between letters, totaling twenty-one columns. The outcome was expressed as the average firing rate of the five columns of a cluster, corresponding to the center letter (O) and a flanking letter (e.g.,

T or N). The value of the afferent inputs to the thalamocortical units representing the center letter O was 38 and the values of the inputs to the thalamocortical units representing the distractor letters T and N were 37 each. The slightly higher value for the inputs at the center site represents the influence of instructions, stored in working-memory structures of the prefrontal cortex (presumed to be the dorsolateral prefrontal cortex; Goldman-Rakic, 1988).

When the three letters are displayed and activity is registered in location maps of the parietal cortex, the prefrontal structure adds one unit to the center site in the parietal cortex to select that location as the target. If the instructions were to identify the first letter of the word TON, then the prefrontal structure would add activation at the coded location of the first letter in the parietal area. The computational goal of the network operations is to magnify the difference between the firing rates of the thalamocortical units between the target and distractor sites, that is, between the five columns representing the letter O and each of the five column sets representing the letters T and N.

Figures 5.10A–C (from LaBerge, Carter, and Brown, 1992) show the firing rates (measured in spike discharges per second) of these two column-clusters over time for the three models, given an afferent firing-rate input of 38 for the center cluster, 37 for each distractor cluster, and 5 for the spaces between. The trajectories of the target

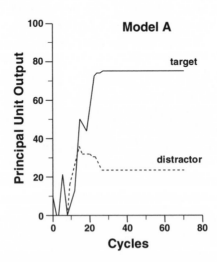

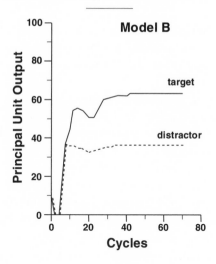

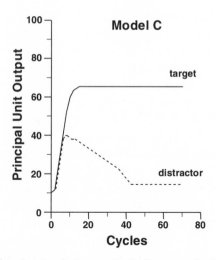

*Figure 5.10.*   Results of a simulation study (LaBerge et al., 1990) of the three versions of the thalamocortical circuit shown in Figure 5.9. Each curve represents the trajectories across time cycles of firing rates of principal cells of a thalamic nucleus. Principal cells corresponding to the target site received an afferent input of 38 units/sec, while principal cells corresponding to the flanker (distractor) site received an afferent input of 37 units/sec.

and surround cells shown in these graphs describe the computations that the set of circuit neurons performs during an interval of time following the arrival of stimulus information in the afferent inputs to the thalamic circuit.

For all three network models of the thalamic circuit, the trajectories of the column-clusters corresponding to the center and surround letters oscillate at first and eventually converge to stable firing levels that differ by more than 25 times the difference at the afferent input. Some settings (not shown here) of the connection weights between neural units induce persisting oscillations of varying amplitude in both curves, and still other settings induce what appear to be chaotic trajectories. The present parameters were arrived at by using a hill-climbing procedure with Model C equations based on the criterion of maximizing the integrated difference between target and distractor firing rates achieved in the shortest time. Most of these parameter values were then inserted into the equations for Models A and B, but a few parameters had to be adjusted by hand to approximate the criterion. Although the asymptotic differences between target and flanker trajectories shown in Figures 5.10A–C are quite large, it can be shown that they are not the maximum differences that the models could produce.

Thus, regardless of the anatomical variations in RN-to-thalamocortical cell projections examined here, the network appears to serve as a mechanism that induces an expression of selective attention in the cortical pathway to which it projects by the enhancement of a target center relative to a distractor surround. The activity at the distractor site shows an initial increase followed by a decrease for all three models (small in Model C), but the comparison of principal-cell activity levels between asymptotic values and initial base rate values (10 spikes/sec) does not yield a negative value for any of the networks. Thus selective attention effects in cortical cells is expressed in all three versions of the thalamocortical network without suppression relative to initial activity at the distractor sites.

The algorithm instantiated by this class of networks apparently can be decomposed into a component that produces enhancement of the target location and a component that laterally inhibits the surround. The enhancement component is produced by the positive feedback of corticothalamic projections, and the lateral inhibition component is produced by the inhibition between the RN units that is projected to the corticothalamic units in neighboring columns. If either component is removed from the network, the output shows little or no magnification of an input target/surround difference. Similar effects were predicted from a neuropsychological theory of attention in cortical circuits described by Walley and Weiden (1973).

Thus, the descending part of the distractor trajectories (following the strong early rise) shown in Figures 5.10A–C reflects a process of active inhibition, not a process of passive decay. The distractor trajectories can be shown to descend to an asymptote of 0, with or without an initial rise, by using parameter values for the network models that are somewhat different from those used to generate the trajectories shown in Figure 5.10A–C. In general, the difference in target and distractor asymptotic outputs will be dominated by an enhancement of the target principal cells when the starting value of all principal cells is low in their range of firing rates (1–100 discharges/sec, here); the difference will be dominated by an inhibition of distractor cells when the starting value of all principal cells is in the upper part of their range.

It is clear that at low initial firing rates of principal cells the network operates on its input chiefly by enhancement of the target-column inputs, with a relatively smaller contribution by the suppression of the surround. However, its effect on cortical cells is not one of inhibition, which means that the circuit cannot suppress the firing rates of cortical cells. Hence, the thalamocortical network acting alone produces selection by enhancement of the target site among cortical cells more than by inhibition of the surrounding sites. This algorithm of attentional expression contrasts with traditional neuro-related theories of attention that have assumed that attention operates mainly to suppress activity at the cortical sites of distracting stimulus input—examples of these theories may be found in Hernandez-Peon (1960, 1966), regarding sensory pathways of the ear; Crick (1984) and Sherman and Koch (1986), regarding the visual pathways at the level of the LGN; and Yingling and Skinner (1977), regarding the reticular nucleus. In the stimulus display TON, in which the target and flankers are closely spaced (within one degree), the pulvinar circuit serves as an enhancement mechanism that produces the expression of attention assumed to occur in the V1-to-IT pathway. The voluntary control that induces the pulvinar circuit mecha-

nism to enhance firing at a particular location within the display
(e.g., at the center of the three-item group) is assumed to originate
in the lateral prefrontal areas, then to project from there to the PPC,
from the PPC to the lateral pulvinar, and next to an area or areas
in the V1-to-IT pathway, such as V4 and possibly TEO.

Figure 5.11 provides a (highly schematic) illustration of this conjectured mechanism of attention operating in the pulvinar between
the PPC and V4. Target and adjacent distractor locations are shown
as coarsely coded by three overlapping distributions in maps of PPC

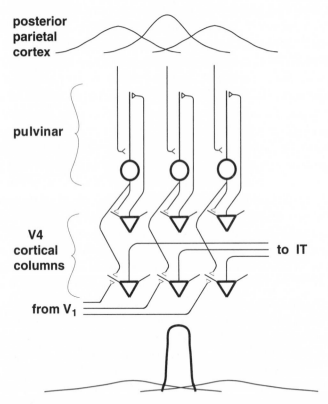

*Figure 5.11.* Schematic diagram of the effects of PPC input on pulvinar output to information flowing from areas V1 through V4 toward IT, based on
simulations that produce both enhancement of the target location (coded by
the center column) and inhibition of the surround (coded by the two outside
columns).

and V4. In the PPC map, the peak of the target distribution is slightly higher than the peaks of the adjacent distractor distributions. In the V4 map, the peak of the target distribution is much higher than the peaks of the adjacent distractor distributions, which reflects the selective enhancement operation of the pulvinar thalamocortical circuitry. The initial values (prior to PPC input) of the principal cells shown in Figure 5.11 are presumed to lie in the lower part of their ranges of firing, so that the output pattern of these cells is generated (according to the model) mostly by enhancement of target principal cells. If initial values at the target and distractor sites were in their upper ranges, then the predicted output peak shown in Figure 5.11 would be generated mostly by suppression of distractor principal cells. But regardless of the initial principal-cell values, the effect of their output pattern is the same, which is to enhance activity at the cortical target site relative to the activity at the cortical distractor sites (as shown in Figure 5.11 for V4). This conclusion is based on the assumption that principal cells project almost exclusively to excitatory (as opposed to inhibitory) cortical cells, as evidence seems to indicate (e.g., Johnson and Burkhalter, 1994).

The present network models have several miscellaneous implications for selective attention. All of the present versions of the thalamic network model adapt to changes in the width of the target area (e.g., LaBerge, 1983), so that similar trajectories for target and surround would be produced if the size of the target area corresponds to the entire word **STONE** instead of just to the letter **O**. Also, the thalamic circuit should be capable of being reset (to some base rate of firing) very rapidly, as in the case of scanning each letter of the word **STONE** from left to right or scanning displayed items in a search task. Present simulation methods show that this can be achieved by setting the corticothalamic firing rate to its base rate, but it is not known what kind of cortico-cortical circuit would project the inhibitory reset signal to the cells corresponding to the target and surround sites. Presumably, the initiation of a reset signal would depend upon the successful completion of the shape identification (e.g., that the center letter is identified as an **O**) in a cortical area that was somewhat remote from the cortical area(s) in which the selective attention is being expressed.

The thalamic network models examined here produce an enhancement in the target/surround difference in the neuromodulatory context of the waking state unless the circuit is restrained by inhibitory influences that are external to the corticothalamic loop. One very noteworthy example of an external inhibitory projection to principal cells in the mediodorsal nucleus and other thalamic nuclei serving the anterior cortex is the projection made by axons arising from cells in the globus pallidus division of the basal ganglia (e.g., Alexander and Crutcher, 1990). These fibers exert a tonic inhibitory restraint on the corticothalamic loop and are themselves inhibited by specific projections from the striatum division of the basal ganglia. The advantage of the tonic inhibition projected to the thalami serving the anterior cortex is that it prevents many cortical regions there from being active simultaneously. In the case of the regions within the premotor and motor areas that control overt movements, the tonic inhibitory restraint on the corresponding corticothalamic loops allows one response to be made at a time. However, the globus pallidus cells apparently do not project to thalamic nuclei whose principal cells project to posterior cortical areas. This suggests that any momentary increase in cortical activity in the temporal lobe, for example, would be immediately magnified and possibly sustained for a period of time. Moreover, many regions could exhibit this elevated activity simultaneously, a pattern of activity resembling a hallucinogenic state. Since normal brain activity apparently does not produce this kind of phenomena, one questions whether corticothalamic projections are alone capable of acting as inputs that result in amplified feedback to the cortical region from which they arise.

In the foregoing section that described the circuitry of the thalamus it was mentioned that the corticothalamic axons terminate on distal dendrites of principal cells. Owing to the distance of the synapse from the soma, these axons may normally function to depolarize the cell membrane but without producing spike discharges. Increased activity at distal dendrites depolarizes the cell membrane closer to threshold value, the consequence being that weaker afferent fiber discharges on proximal dendrites can induce the principal cell to fire. This implies that corticothalamic activity alone should not induce firing in principal cells, at least during the normal waking state, and the generation of hallucination-like states in the posterior cortex by means of the cortico-thalamo-cortical loop may be ruled out. To account for the new assumption concerning distal dendritic synapses, the present thalamic network models would have to be modified to allow distal dendritic activity to control the momentary connection weights of afferent synapses instead of assuming these weights to be constant.

The present description of how the thalamic circuit could instantiate the algorithm of selective attention by increasing target/surround contrast in firing rates has assumed that targets and flankers (distractors) are sufficiently close (in space and/or in attribute dimension) that the reticular nucleus cells can produce lateral inhibition between the principal cells of the target sites and the distractor sites. If the target and distractor sites are sufficiently separated (e.g., in opposite visual hemifields), then the thalamic network described here will not produce a contrast in activity between target and distractor sites, owing to the lack of connections between the RN cells. Instead, the thalamic columns corresponding to the target and distractor sites will separately enhance the afferent input activity, treating each stimulus as if it were the only object in the visual field. For example, a dot appearing in each hemifield would induce the circuit to produce a small enhancement, but when attention is directed to the site of one of the dots by the addition of activity to the corresponding thalamic input, then the thalamic circuit could enhance this top-down input and thereby magnify the output difference between target and distractor sites. Thus, the enhancement effect of the thalamic circuit can operate regardless of the separation of a target and its (nearest) distractor. As the distance between target and distractor (in space or attribute dimension) increases, however, the amount of target enhancement needed to respond appropriately to the target will decrease.

When target and distractor are sufficiently close to be initially processed as one object, then there is no ambiguity in the current eye-movement information; but when target and distractor are separated sufficiently to be sensed as different objects, then conflicting eye-movement information may be present. Eye-movement information, generated in mechanisms of the superior colliculus and the posterior parietal area, may be projected through thalamocortical circuits to cortical pathways, where attention is expressed in the kinds of object-identification tasks being considered here.

Area LIP (lateral intraparietal), within the PPC, has been implicated strongly in processing eye-movement information, and its cells project not only to the SC, which generates commands to the brainstem control centers, but also to V4 and TEO (Blatt et al., 1990; Nakamura et al., 1992). The location of objects in a display are presumably coded, though coarsely, in both LIP and DLPFC maps, which are directly interconnected (Goldman-Rakic et al., 1993). When subjects voluntarily (e.g., in response to instructions) choose to attend to a particular location, the location of the chosen object is presumably mapped in the DLPFC and projected to the corresponding location in the LIP map (and perhaps other maps in PPC as well). The activated location in the LIP map then is projected to a corresponding location in V4 and TEO maps, resulting in an enhancement of activity at the location of the target site while activity at the distractor site remains unchanged or, possibly, is enhanced slightly. The projection between LIP and V4, for example, would also be accompanied by a projection through the lateral pulvinar, resulting in some enhancement at both target and distractor locations, owing to the wide separation of the two objects. If the target object required a difficult attribute discrimination (as in the tasks required by Spitzer et al., 1988, and Haenny and Schiller, 1988), then additional enhancement would be predicted at the target location. According to this account, therefore, the expression of selective attention in V4 and TEO for highly separated target and distractor(s) is one of enhancement at the target site, which contrasts with the suppression-only expression of selective attention indicated by Moran and Desimone (1985).

*The thalamus as a mechanism of selective attention.* This section of the chapter has examined the thalamus anatomically in terms of its internal circuitry and connectivity to other brain areas, and functionally in terms of its activity during attention tasks as indicated by physiological measures and by simulations based on network models of its internal circuitry. These considerations appear to converge on the hypothesis that the thalamic circuitry can produce an enhancement of activity in selected sites of cortex and thus modulate flow in cortical pathways. The PET experiments reported earlier (LaBerge and Buchsbaum, 1990; Liotti, Fox, and LaBerge, 1994) employ task displays in which distractors of high similarity to the target are placed very close to the target. Attending to the location of the target in these tasks appears to induce a high concentration of attention, as indicated by levels of blood flow observed in the pulvinar and mediodorsal nucleus (in Liotti, Fox, and LaBerge,

1994) of the thalamus. The sources of control on these thalamic nuclei appear to be the prefrontal cortical areas, where voluntary processes initiate the selection process with projections of activity that are enhanced by the thalamus en route to the expression of attention in cortical pathways. When distinctions are very close, attentional activity presumably rises to very high levels, which may prompt subjects to report that the attended content "fills" or "possesses" the mind. High levels of attentional enhancement also protect actions from the interference of other concurrent actions, and protect sustained perceptual processing from the potentially powerful interference by sudden onsets of environmental stimuli (Bacon and Egeth, 1994; Theeuwes, 1991, 1992; Yantis and Jonides, 1990). Apparently, internal sources of attentional control are able to enhance cortical processing to levels that are higher than those induced by strong onsets of external stimuli.

High levels of "mindfulness," presumed to be induced and sustained top-down from sources in the prefrontal cortex, may place the individual at a disadvantage under some circumstances. When attention is directed strongly to one object, idea, or plan of action, the individual may fail to notice an impending threat from some part of the environment. The process by which internal, endogenous control of attention attenuates the effects of exogenous influences on attention has been discribed in a recent review of the relevant literature by Yantis (1993).

*Comparing modulatory "emphasis" in the retina and the LGN, pulvinar, and mediodorsal nuclei of the thalamus.* The visual system, as well as other sensory systems, appears to apply modulatory operations on incoming information at several stages as the information flows toward the higher levels of the brain, where it encounters processes of voluntary action. In the retina, light energy is transduced into neural activity at the rods and cones, and the horizontal cells that link these photoreceptors exert lateral inhibitory effects that increase the contrast between boundaries of lumination. This increase in contrast can be regarded as a modulatory "emphasis" of the information at the boundary relative to information between boundaries. Lateral inhibitory connections at the next retinal level consist of amacrine cells that link bipolar cells, and this stage of early retinal processing is also believed to increase the contrast in firing rates at luminance boundaries.

The ganglion cells at the third layer of the retina project their axons to the thalamocortical neurons of the LGN, where "emphasis" of contrasts appears to take place. But the modulatory emphasis of contrasts in the LGN is accomplished not simply by lateral inhibitory links between pathways of information flow but by a circuit in which lateral inhibition is combined with recurrent excitation. The structure of the LGN circuit has the "standard" form of the thalamocortical circuit, and three slightly different versions of this circuit, shown in Figure 5.9, were examined in some detail in previous sections of this chapter. The lateral inhibitory connections in this circuit occur between cells of the reticular nucleus (often called the perigeniculate nucleus when it serves the LGN), and the recurrent excitatory connections occur between cortical cells (mainly in layer VI) and thalamocortical cells. Simulation results, described in this chapter, suggest that the recurrent excitatory connections produce greater contrasts in activity between optical pathways than are produced by lateral inhibition alone. But lateral inhibition is apparently required to allow recurrent excitation to produce these greater contrasts. Thus, the thalamocortical circuitry of the LGN provides a means of increasing certain kinds of contrasts at the boundaries of visual objects by adding an enhancement operation to the lateral inhibition operation.

Thus far, the modulatory emphasis exerted on particular aspects of visual input appears to be built into the hardware and to occur automatically in a bottom-up manner whenever stimulus objects are displayed to the individual. But in many natural visual scenes many objects occur, and each object typically exhibits several attributes, with the result that the retina and LGN produce many objects and attributes having relatively equal emphasis. If the individual is to act appropriately in the face of equally emphasized objects, one object must be at least momentarily emphasized further. The type of modulatory circuitry of the LGN would seem adequate to the job of further emphasis, but it needs the added property of receiving controlling inputs from anterior cortical areas where actions and motivation are processed, so that modulatory emphasis is exerted on the adaptively appropriate object or attribute.

The added property of top-down control on modulatory emphasis is given to other thalamic nuclei, such as the pulvinar (and presumably the lateral posterior nuclei in lower animals who possess no pulvinar). The input to the pulvinar from the anterior cortical areas is apparently indirect: it first is registered in extrastriate areas of the posterior cortex, for example in the posterior parietal and temporal areas, where it presumably encounters visual information from area V1 that has already undergone modulatory emphasis by the retina and LGN. For example, the selection by higher levels of the middle object of a display of three identical objects (e.g., three apples) arranged in a row adds activity to that object's representation in the parietal map, where the three object locations are initally coded from bottom-up information in area V1. The information from the three objects in the parietal map is then projected into the pulvinar, where thalamocortical circuits modulate information flow to emphasize the activity of the middle object over the activity of its neighbors.

But the higher-order processes concerned with constructing, maintaining, and executing action plans are themselves subject to emphasis. Again, the thalamocortical circuit, with its combination of recurrent excitation and lateral inhibition, appears to have the ability to provide a large range of modulatory emphasis that is adequate to the job of processing action information. The anterior cortical areas are known to be crucially involved in processing action information, and the main thalamic nucleus that is most strongly connected to these areas is the mediodorsal nucleus.

Thus, the notion of modulatory emphasis can be used to relate modulatory processing with neural structures from the retina to the higher levels of the brain, where action plans and execution take place. Table 5.1 provides a summary of the conjectured relationship between each mechanism's modulatory process and the neural structure containing the mechanism.

*Summary.* This section on the thalamus describes a variety of anatomical, physiological, and simulation results that appear to converge on the conclusion that a mechanism of selective attention is embodied in the thalamocortical component of the triangular circuit connecting a pair of cortical areas. For highly distinct locations and attributes, the direct route between two cortices may suffice to produce the expression of attention to a particular location or attribute; for less distinct locations and attributes, the indirect thalamic route between two cortices is presumed to be required to resolve information flow and produce the expression of attention to a location or attribute. The superior colliculus may contain an appropriate

*Table 5.1.*   Mechanisms of modulatory "emphasis"

| Structure | Mechanism(s) |
| --- | --- |
| Retina | Lateral inhibition |
| Lateral geniculate nucleus | Lateral inhibition plus recurrent excitation |
| Pulvinar nucleus | Lateral inhibition plus recurrent excitation under top-down control acting on areas of perception |
| Mediodorsal nucleus | Lateral inhibition plus recurrent excitation acting on areas of action |

mechanism for expressing attention to highly distinct locations of objects in the posterior parietal areas, although the outputs from the SC pass through thalamocortical circuits that may provide further attentional modulation before the SC outputs reach a cortical area. The superior colliculus is assumed also to be involved in registering new information sources at their sudden onsets, which is termed "noticing" here. When objects are clustered together the thalamocortical mechanism is assumed to resolve the information arising from one particular object, but even when an object occurs in isolation the thalamocortical circuit can enhance the information arising from that object within cortical pathways. These early processes of noticing, selecting, and enhancing object information shape that information for subsequent processing of object identity, classification, color, velocity, and the like, which is termed "realizing" something about the object. The noticing-selecting-enhancing-realizing sequence of events is not confined to the perception of external stimuli but applies also to ideas of objects and actions that are noticed as they appear in memory. Future research of brain activity during attentional processing in perceptual and ideational tasks will continue to illuminate the interactive functions of cortical areas in which these processing events occur, as well as the modulatory effects of thalamocortical circuits on these interactive cortical functions.

# 6

A Cognitive-Neuroscience
Model of Attention Processes
in Shape Identification

This chapter provides a condensed synthesis of the material presented so far. Chapters 1–3 treated attentional processing at cognitive levels of description, Chapters 4 and 5 at neurobiological levels
of descriptions. Cognitive concepts of attention, such as activity distributions of preparation for locating a target or for organizational
anchoring in problem-solving, are relatively global, in comparison
with detailed neurobiological descriptions of enhanced firing of
cells in a cortical pathway. But an adequate account of attentional
processing requires a wider scope than a description of activity in a
group of local cortical cells, regardless of how detailed that description may be. Earlier chapters stressed the point that attentional effects occur simultaneously in many areas of the brain, cortical and
subcortical, and that it is important that the expression of attention
in cortical pathways be distinguished from the mechanisms that produce these expressions and from the brain structures that exert control over those mechanisms. Thus, an adequate neurobiological account of attention must incorporate descriptions of circuitry that
may extend across many architectural areas of the brain. In order
to understand how these far-flung components of brain architecture
are coordinated as an individual is engaged in performing a task,
however, it would seem necessary to draw on the kinds of global
descriptions of attentional processing provided by cognitive psychology.

## An Experimental Trial Containing
## a Warning Signal and a Target

In view of these considerations, this chapter presents a combination of cognitive and neuroscience descriptions of attentional processes as these processes are presumed to occur within a trial of an experiment. The particular experiment to be analyzed here contains trials that consist of two displays, presented one after the other: a warning signal, ###, and a target, QOG. The subject is asked to identify the center letter of the target display by pressing a button when the center letter is a O. To insure that an identification process actually takes place, half the time the letter O is replaced by the letter C or the digit Ø. The locations of the G and Q distractors was randomized. This type of target-distractor display was used to intensify attention in the two PET studies described in Chapter 5 (LaBerge and Buchsbaum, 1990; Liotti, Fox, and LaBerge, 1994), and also in the response-time experiments that produced the data shown in Figures 3.2, 3.3, and 3.4.

Figure 6.1 extracts the relevant anatomical data from Chapters 4 and 5 to show the main brain areas involved in attentional processing of the visual shape along with their principal interconnections. Figure 6.2 depicts the cognitive type of information flowing through these brain areas while the warning signal of the trial (###) is displayed, and Figure 6.3 depicts cognitive information flowing through these brain areas while the target (QOG) is displayed.

*Attentional processing during the warning-signal display.* Warning sig-

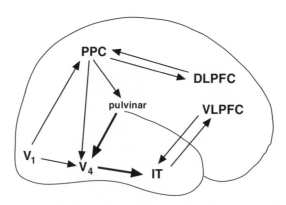

*Figure 6.1.*   Major brain areas involved in attending to a visual object.

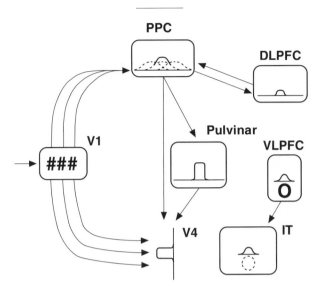

*Figure 6.2.* Schematic model of preparatory attention at the end of a warning-signal display period. The subject is prepared to process a particular shape, O, at the center location of the warning signal.

nals customarily prepare a subject for the stimulus events that follow. As the duration of the warning signal approaches its termination, attentional preparation for the target stimulus which is about to appear intensifies. In the present examples of a trial in a shape-identification experiment, the duration of the warning-signal interval is on the order of a second, which provides sufficient time for the subject to build preparatory attention for the target's shape and location to a high level of intensity.

The warning signal is displayed as three pound signs (###) at the center of the screen and remains there for 1,000 msec. At the onset of the warning signal the information flows in two major directions: upward (dorsal), toward the location maps of the posterior parietal cortex (PPC), and downward (ventral), toward the identification modules in the inferior temporal cortex (IT). The warning-signal information is projected onto the location map in the PPC, which codes the location of the warning signal in a somewhat coarse manner, owing to the large receptive fields in this cortical area. The location of the objects in the PPC map is then projected onto a map in

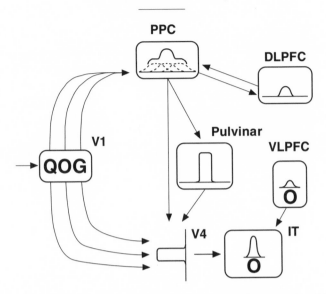

*Figure 6.3.* Schematic model of preparatory and selective attention following the onset of the target display. Preparatory attention to both object and location is sustained from the warning-signal period, and additional inputs from the stimulus (registered in V1) and from the memory of the instructed target location (stored in DLPFC) raise the corresponding activation levels in PPC. Activation from DLPFC continues until the activity distribution in V4 becomes sufficiently concentrated and sustained long enough to deliver the unambiguous information needed by IT to identify the center letter of the stimulus display.

the dorsolateral prefrontal cortex (DLPFC), where it is maintained in working memory along with the memory of instructions to select the object in the upcoming target display that is located at the same position as the warning signal.

The coded location of the warning signal in the DLPFC may be used by the subject to induce an attentional preparation for the spatial location of the upcoming target letter, which in this experiment is registered in the PPC at the location of the center object of the warning signal. The subject may or may not choose to direct activity from the location code in the DLPFC to the PPC, where it can increase the activity at the center of the activity distribution there. The existence of this option underscores the assumption that working memory, as viewed in this book, contains two separate components related to attentional preparation: the expectancy of a particular object occurring at a particular time, and the action command to induce an expression of preparatory attention in a posterior cortical map corresponding to the location and/or attribute of the expected object (see the discussion of the distinction between expectancy and preparation in Chapter 3).

During the warning signal, the subject elects to prepare perceptually for the upcoming target at the center of the warning signal's pound signs, as shown in Figure 6.2. The intensity of attentional preparation in this example is moderate; consequently the activity distribution in the PPC shows a moderate rise at its center. The dotted-line distributions represent the activity produced from the sensory input of the warning signal plus any residual activity at locations of stimuli presented in previous trials (see Chapter 3's discussion on the activity distribution of preparation).

The entire distribution of activity in the PPC is then projected through the pulvinar, which enhances this activity at the highest point (mode) of the distribution and reduces it at surrounding locations. When this transformed activity from the PPC is projected to the V1-to-IT pathway, it is assumed to be added to the activity distribution already flowing from area V1. The added activation to the pathways from V1 containing the shape information of the pound signs, if focused sufficiently on the center shape, may induce an identification of that shape. In the case illustrated in Figure 6.2 the pound sign may have already been identified by the enhanced flow of information in the center pathways of the flow of information from V1 to IT. No identification of the central pound sign is shown in Figure 6.2, because at the end of the time the warning signal is displayed the identification module is assumed to express preparatory attention for the upcoming target letter O.

The preparatory attention for the expected target shape O is assumed to be generated from the ventrolateral prefrontal cortex (VLPFC), a working-memory area that lies just below the DLPFC area that maintains the code for the expected location of the target. As in the case of the DLPFC, expectation of the letter O in the VLPFC (which may be in verbal form) is separated from the action commands that project activity from the letter code to its perceptual form represented in area IT. Hence an expectation for the letter O may exist without associated actions. In the present example, the subject chooses to attend in a preparatory manner to see the letter O, and the preparation process may be regarded as a form of visual imagery of that shape (Kosslyn, 1994) that dominates IT processing while pound signs of the warning signal dominate processing in early visual pathways (e.g., in V1 and V2).

The expectation for the letter O together with its expected location are presumed to be conjoined in an appropriate code. This superordinate code, held in working memory, resembles an "object file," described by Kahneman, Treisman, and Gibbs (1992), which coordinates the directing of attention to the spatial and shape features of that object in sensory maps.

*Attentional processing during the target display.* When the warning signal goes off and the target immediately appears, much of the processing that occurred during the warning signal carries over into the processing of the target. Figure 6.3 shows a "snapshot" of attentional activity in the most relevant brain areas during the first moments of the target display in the same fashion as Figure 6.2 showed a "snapshot" of attentional activity in the same brain areas during the later moments of the warning-signal display.

When the target display appears, the information arising from the entire QOG stimulus flows in the dorsal and ventral directions. In the PPC the information flowing from the V1 registration of the three-letter stimulus increases activity at all three letter locations over the preparatory attentional activity remaining from the projection from the DLPFC during the warning period. The preparatory activity held over from the warning-signal period in the PPC (generated from the DLPFC if the subject chose to produce perceptual preparation from the stored expectancy of the target object's location) is important because it is one of the two sources of modulation that ultimately produce faster identifications of the target letter. The other source of perceptual modulation is the preparatory activity for the shape shown within the IT module of Figure 6.2.

The information from the entire QOG ensemble is also relayed to the DLPFC map, where it is coded in terms of its three objects. On the basis of stored instructions to select the center object, the DLPFC returns activation to the PPC at the center object location, which is represented by the elevated peak at the center of the PPC distribution in Figure 6.3. When this total distribution is projected to the pulvinar, the circuitry of the pulvinar enhances activity in the pathways corresponding to the modal location and reduces activity in surrounding pathways. The combined enhancement/decrement operations results in a considerable sharpening of the distribution of activity across locations as it is projected onto the V1-to-IT pathway.

The role of selective enhancement by the pulvinar in selecting information at the target site in the V1-to-IT pathway may be more clearly seen by considering the effect of the direct connection alone between the PPC and the V1-to-IT pathway. If the activity distribution in the PPC (in Figure 6.3) were projected directly to the V1-to-IT pathway, there would be little modulatory difference induced in the information flow at the target and distractor locations. Hence the information entering the identification modules in IT would arise from all three letters instead of only from the center letter. If the target display were COG or HOT (which have memory codes in IT, while QOG does not), then the identification module should identify the three-letter ensemble as the word *cog* or *hot* and not the center letter *O*. Thus, if information from only the target location is to enter the identification modules, the activity level of that information must be considerably greater than the activity levels of information arising from the adjacent distractors. The thalamocortical circuitry involving the pulvinar is assumed to produce the required modulation of information flowing from the PPC to the V1-to-IT pathway.

It is clear that the model, shown schematically in Figures 6.2 and 6.3, allows for the possibility of identifying objects without the selective enhancement operations of the pulvinar. For example, displays of a single O (i.e., without distractors) generate a flow of shape information from V1 to IT that requires no modulation in order to produce unambiguous information necessary for the identification of the letter in IT because the activity in the target pathways may already be well above the activity in the surround. In like manner, targets presented with distractors at several degrees' distance from the target may need no modulation of V1-to-IT information flow for their correct identification. Also, targets presented with dissimilar distractors nearby may be correctly identified without attentional modulation prior to their entering the IT modules, owing to prior enhancements of feature contrasts produced by the LGN thalamo-cortical circuitry and triangular circuits at areas V2 and V3 (see Chapter 5) before the information arrives at area V4. But as adjacent distractors become more similar to the target, the activity levels in the target and distractor pathways leaving V1 approach equality, so that modulation of the V1-to-IT information flow becomes necessary to produce the required contrast in target/surround activity and to enable just the information from the target to enter IT and produce a correct identification.

Attentional enhancement of the location information of the target within the V1-to-IT pathway can speed the identification of the target object even when the distractors are dissimilar or positioned at a distance or even when distractors are absent. Thus, it could be said that attentional enhancement always aids perceptual processing but is not required for selective processing unless distractors similar to the target are positioned close to the target.

*Targets appearing at unexpected locations.* If the QOG target, shown in Figure 6.3, were presented off to the left or right of the expected and prepared-for location, how does the present model predict that time to identify the target will increase (the diagram of preparatory attention in Figure 3.1 is helpful here)? When the QOG stimulus is presented, say, a few character spaces to the left of center, information flow from V1 and the DLPFC will increase the height of the activity distribution in the PPC at the coded location of the QOG stimulus (on the left side). But the existing height at the center location will compete with the height at this left-side location for dominance as the PPC activity is projected through the pulvinar. In order to build a new mode at the location of the target O in its unexpected location to the left of center, the DLPFC must project more activation to this new location. This additional projection from the DLPFC requires more time than is needed when the QOG stimulus falls at the expected location, that is, at the center mode of the distribution in the PPC produced during the warning signal. If the QOG stimulus is presented further from center, the time that the DLPFC must project additional activation to the PPC would be even greater. Thus, response time to the target will monotonically increase with distance from the expected location in the center, as shown by all of the curves of Figures 3.2–3.4.

## Summary: The Expression, Mechanism, and Control of Attention in Shape Identification

In Figures 6.2 and 6.3 the preparatory attention to the spatial location of shape information is assumed to be expressed in the V1-to-

IT pathway by an enhancement of target information relative to distractor information. Even in the absence of distractor information, the spatial location of target information may be enhanced to speed the identification process. The mechanism that produces this enhancement is assumed to be the thalamocortical circuits of the pulvinar, and the control of the location, width, and intensity of the enhancement is assumed to involve memory components and retrieval components within the DLPFC. Preparatory attention to location is also expressed in the PPC (shown as the thickest lines of the PPC activity distribution in Figure 6.2). The mechanism that is assumed to influence the enhancement in the PPC (via a triangular circuit) is the mediodorsal nucleus of the thalamus (not shown here), which is known to receive fibers from the DLPFC and to send fibers to the PPC (Siwek and Pandya, 1991; Schahmann and Pandya, 1990). This route through the thalamocortical circuitry enables DLPFC projections to control the attentional enhancement at the precise location of the center target in the PPC map. Thus, aligning preparatory attention at a precise location in the PPC is also presumed to involve a thalamic selective enhancement mechanism.

Preparatory attention to the shape of the expected target is expressed within the identification module of area IT, as shown in Figure 6.2. The mechanism that can enhance this expression to varying degrees of intensity is assumed to be a thalamocortical circuit (in a different part of the mediodorsal nucleus than the part involved with DLPFC projections), and the control of the mechanism and the expression of attention in IT is assumed to lie in the VLPFC (ventrolateral prefrontal cortical region). As in the case of the DLPFC, this cortical area is assumed to contain a working-memory component (Chelazzi et al., 1993; Wilson et al., 1993) and a retrieval component.

While selective enhancement operates over relatively extended time periods of preparatory attention during a warning signal, selective enhancement operates rather quickly when the target appears and is identified. In Figure 6.3, the expression of attention to spatial location in the V1-to-IT pathway need only be long enough to enable the IT module to identify the target shape, perhaps on the order of 20–50 msec (Tovee et al., 1993). In like manner, the operation of the pulvinar mechanism and the control processes in the DLPFC would seem to be quite short in duration as well (unless the target stimulus is located quite far from the expected location). Within the line of control is the expression of selective attention to location in the PPC, which itself is assumed to be shaped by the thalamocortical mechanisms of the mediodorsal nucleus. Finally, it is conjectured that, when the target O is identified in area IT and this event projects signals to prefrontal areas in which actions are processed, the VLPFC area would be one of the areas receiving this projection and this area would again induce thalamocortical circuits in the VLPFC to enhance selectively this activity as it flows toward areas that generate actions.

The analysis of shape identification in this chapter has singled out four different brain areas on which attention is expressed, and several of these expressions are assumed to take place simultaneously. Each area exhibiting attention is connected to a thalamic nucleus in which thalamocortical circuits serve as local mechanisms that produce the corresponding elevated activity patterns in their cortical areas.

Identifying a visual object is only one of the many perceptual judgments carried out in daily life. Colors, sizes, movement directions, and velocities of visual objects are other visual judgments that may require selective enhancements of a location and/or an enhancement of the attribute involved. Perceptual judgments of auditory, tactual, olfactory, and gustatory stimuli often require selective emphasis, and thalamocortical circuits connected to the cortical areas involved in these judgments are presumed to carry out this function. The analysis of shape identification in this chapter may suggest ways to approach the particular ways that preparatory and selective attention operate during the processing of these judgments.

The selection of action codes during the planning of responses (both internal ideational acts and external motor responses) typically involves both a preparatory phase and an execution phase, and the operation of selective enhancement described here may suggest ways to analyze these cognitive activities as well.

The basic concepts that underlie these and other examples of attention-assisted cognitive processing are: (1) the way attention is assumed to be expressed in cortical pathways; (2) the mechanism most directly responsible for generating the form of this expression; and (3) the brain areas that control this mechanism. Since virtually every cortical area appears to be connected to a thalamic nucleus in the manner given by the "standard" thalamocortical circuit (e.g., as shown in Figure 5.6), it seems likely that, given the evidence and theoretical considerations reviewed in the foregoing chapters, virtually every cortical area is capable of undergoing at least some degree of some selective enhancement from its thalamocortical circuit. Thus, the concepts of attentional expression and its associated mechanism provide a unifying base for understanding attention as it operates during the myriad forms of cognitive processing.

# 7

# Synopsis

In this book attention has been examined from several perspectives, including those arising from the study of cognition, behavior, computation, and neurobiology. An attempt has been made to show how certain approaches taken from each of these fields intersect to produce a particular cognitive-neuroscience view of attention. What follows in this concluding chapter are four general premises and a summary of this cognitive-neuroscience view of attention.

(1) *Attention is biological; it is not a culturally based processing program acquired by environmental training.* The ability to attend is built into the "neurological hardware," just as respiration and circulation are built into other bodily tissues. It is assumed that the algorithmic form of attentional expression in brain pathways is almost entirely determined by the underlying hardware. This view contrasts with current theories regarding the form of expression of cognitive skills, such as subtracting numbers, reading, or the acquiring of a specific database. Cognitive skills like these can be carried out by different computational procedures, and their algorithms depend strongly on how the individual was trained. The plasticity of attentional processing exhibited in the process of learning virtually all skills, cognitive and motor, is attributed mainly to plasticity in neural structures that regulate or control the mechanisms of attentional expression. The plasticity that may exist in the mechanisms of attention is assumed to consist in adjustments to scalar values of the algorithm, not in changes to the procedural form of the algorithm.

(2) *Discovery of an algorithm of attentional expression requires computa-*

*tional, cognitive, and behavioral knowledge as well as knowledge of the anatomy and physiology of neurons and neural circuits.* If one is to understand how neural tissue might express attentional processing, one must first have a conception of what attentional processing does—that is, what is computed when attentional processing takes place. (This point was made with exceptional clarity for the case of visual processing by Marr, 1982.) This conceptualization can then guide the design of appropriate behavioral tasks, and it may point to ways of directly stimulating cells to induce the neural tissue to function in an attentional mode.

(3) *The expression of attention may be manifested to human observers by a variety of indicators: behavioral, physiological, and experiential.* Behavioral indicators are separated from the event of attentional expression by several inferential steps. In contrast, physiological indicators, particularly single-cell recordings, are more directly related to attentional expression. Experiential indicators generally have a relatively close proximity to the expressive event, but these are subjective rather than objective indicators and are of only limited usefulness for theoretical formulations. The main behavioral indicators of attention in the research described here are: the correct identification of items when other items also appear in the visual field (e.g., identifying the letter O in the display HOT) and the speeded correct response to an anticipated item, attribute, or location, with and without other items in the visual field (e.g., identifying a cued red dot against an alternatively presented orange dot). The first case manifests the selective expression of attention, which is particularly evident when items are spatially positioned close to the target items. The second measure manifests the preparatory expression of attention in conjunction with its selective expression.

The most frequently used physiological measures are: ERPs, whose wave components exhibit amplitude differences over brain areas in which attention is expressed in suitably designed tasks; PET images, which reveal blood-flow differences or differences in glucose uptake in brain areas in which attention is expressed during suitably designed tasks; recordings of differences in discharge rates of specific cells in areas in which attention is expressed during appropriately designed tasks. Physiological correlates of behavioral measures are sites of brain lesions in patients showing deficits in attention tasks.

Experiential manifestations may offer the human observer the most vivid and rich indicators that attention is occurring, but these indicators do not generally provide an appropriate base from which to infer precise properties of attentional processing. Yet, experiential observation of attentional activity has traditionally served as a highly useful guide for theory and research. William James, whose experiential observations of the workings of attention are still without equal, noted "the failure of psychology in general to uncover a single elementary law" (quoted in Perry, 1967)—and this held true for the area of attention as well. At the same time James claimed (1890, vol. 1, p. 402) that "my experience is what I agree to attend to." This situation—having a strong general idea without the practical details—parallels a similar episode in the history of chemistry, when enzymes were known to be important but no progress was being made to determine how an enzyme operates to speed interactions of other chemicals. Thus, although experiments carried out during James's lifetime fell short of the goal of reaching a principled understanding of attention, the private observations of his own mental activities led James to regard attention as having a central role in the stream of cognitive processing that fills our waking hours. The frequent references in the research literature to the notions put forth by William James attests to the heuristic value of hypotheses arrived at through one's own experiences.

Although personal experience may instruct an individual about both the selective and the preparatory properties of attention, it arguably provides the most salient evidence we have of the maintenance of attention to an external stimulus or to an internal idea. Here the goal of the attentional computations is the continued processing of the attended content, whether that be the flavor of an excellent wine or the contemplation of an intriguing idea; however, the survival value of the continued attending is not as apparent as the survival value of attending to objects and their properties for the purpose of correctly and quickly identifying them in order to make prompt adaptive responses. Nonetheless, the prospect of extended durations of attention provides much of the motivation for actions we take in our daily lives, and this fact warrants additional research effort into this manifestation of attention so that we may come to understand it as well as we do the selective and preparatory manifestations of attention.

(4) *The expression of attention is not restricted to one brain area but can*

*occur in many areas, and it can take place in more than one area simultaneously (depending upon the areas).* However, the target of attention is apparently restricted to one entity within the same functionally defined area (e.g., attending to a specific color in a specific location), except possibly very briefly under special circumstances (e.g., attempting to attend to two colors or two locations simultaneously). The implication of this premise is that there are anatomically separate circuit mechanisms for the expression of attention in each brain area that are under independent control, but that circuit mechanisms within an area are not functionally independent owing to interactions within the controlling structures and/or between the circuit mechanisms serving the area.

## A Cognitive-Neuroscience Theory of Attention

Attention is defined in terms of its expression in brain pathways. Attentional expression is a positive difference between information flow at the attended site and the information flow in its surround. The information flow that undergoes attentional modulation may arise from sensory sources or from memory sources. The expression of attention may vary in strength from a low intensity (of concentration) to a high intensity and may vary in duration from relatively brief or transient to relatively long or sustained. When attentional expression is sustained at a high concentration, it is experienced most vividly and is said to "possess" the mind or "fill" the mind.

Computationally speaking, the selective use of attention does not require an enhancement of target information over some base rate of information flow; it requires only that the target information flow be greater than the information flow in the surround or at distractor sites. This target/surround difference can be realized in three ways: an increase in output at the target subset that is greater than the increase of the non-target subset (if it increases at all); a decrease in output at the non-target subset that is less than a decrease in output at the target subset (if it decreases at all); or an increase in the output at the target subset and a decrease in the output at the non-target subset. If the output difference is produced by a decrease in the output at the non-target subset, then selection can occur without enhancement of the target subset. When base rates of firing in cortical pathways are low, however, then large differences in information flow between the attended site and its surround would have to be produced mainly by an enhancement of the flow at the attended site.

It is assumed that attention may be expressed in several different brain areas simultaneously, as when an individual attends simultaneously to the color, shape, and location of an object. Within a brain area, high concentrations of attention may be limited to processes that "realize" one attribute or location at a time when attention is controlled by anterior cortical areas; when attention is controlled from sensory input sources, more than one value of an attribute or location may be momentarily enhanced; while these enhancements could be regarded as multiple (although very brief) expressions of attention, this book refers to these enhancements as the "noticing" of sources of sensory inputs (prior to their being "realized"). Attentional expressions are assumed also to occur within the anterior cortical areas, such as those that process plans of internal and external actions.

*What does attention do for the individual?* The site in brain pathways where attention is expressed is the site where attention produces its effects on the information flow. The consequence of effective attentional expression in various brain pathways is a modulation of neural information processing that promotes the survival of the organism and prolongs desirable cognitive states. Thus the two general types of goals served by attention are behavioral and experiential. Examples of behavior goals are appropriate and prompt actions based on correct and rapid judgments of objects and attributes in the environment. Correct judgments are served by the selective property of attention, which reduces the ambiguity inherent in the information arising from an array of closely spaced similar objects. Planning an appropriate action sequence is also served by the selective property of attention, as when a particular goal image provides an organizing anchor for execution of the various operations that make up an action sequence.

Along with the selective benefits of attention are the benefits of enhancing activity levels of cognitive operations, which may be said to include preparation, protection, and prolongation. Preparatory attention promotes rapid processing by increasing the activity in particular pathways prior to the onset of an expected stimulus or an anticipated response. Enhancement of the activity levels of particular cognitive operations protects them from the interfering influences of other ongoing operations and especially protects them from the disrupting effects of sudden onsets of distracting environmental events (e.g., more attentional concentration is usually required when one is reading a book amid loud traffic noises than when one is reading in silence). Experientially, attention can serve to prolong or maintain mental states that are usually pleasurable but may also be unpleasant, such as feeling anxious or worried. If these affective mental states are enhanced to relatively high levels, the experienced pleasures or pains may be commensurately increased. Thus, potentiation (of affective states) may possibly be added to the other three enhancement benefits of preparation, protection, and prolongation.

*The mechanisms of attention.* The mechanisms of attention directly modulate the flow of information in cortical pathways that constitutes the expression of attention. In the case of visual spatial attention, which has received the most extensive treatment in this book, the two dominant mechanisms considered are neural circuits between the thalamus and cortical areas, and neural circuits involving the oculomotor regions of the superior colliculus and the posterior parietal cortex. The algorithm embodied in the thalamic mechanism, suggested by simulations of the thalamocortical circuitry, produces a difference in information flow at the target and surround locations (attended and non-attended sites) mainly by the enhancement of flow in brain pathways corresponding to the target site, whereas the algorithm embodied in the oculomotor-related mechanisms is conjectured to produce a difference in target/surround information flow mainly by the suppression of flow in brain pathways corresponding to the surrounding sites.

Because base rates of firing in most cortical pathways are low, the production of large differences in target/surround flow must be accomplished by enhancements of flow at the target site instead of by suppressions of flow at the surrounding sites. Therefore, high concentrations of attention in these areas are assumed to involve the thalamic mechanism. On the other hand, low concentrations of attention in these areas may be produced either by enhancing the flow mostly at the target site or by decreasing the flow mostly at the surrounding sites or by a combination of the two operations.

*The control of attention.* The mechanisms of attention do not serve as the cause of attentional expressions in cortical pathways. Rather, the causal determinants of what will be attended to, the intensity of attention, and the duration of attention lie mainly in areas of the prefrontal cortex that embody voluntary processes. It could be argued that very brief attentional expressions in cortical pathways may be produced by the sudden onsets of intense stimuli, but these short-lived surges of activity in cortical pathways are viewed in this book as instances of noticing the information source rather than attending to the location and attributes of the objects or events related to this source.

Anterior cortical areas that physiological measures have identified as controlling attention are the dorsolateral prefrontal, ventrolateral prefrontal, and anterior cingulate cortices. The dorsolateral and ventrolateral areas have been shown to contain neurons that specialize in spatial location and in attributes such as color and shape, and these neurons discharge during delays between a cue and a target, suggesting that they serve as working memory for location and shape attributes. The anterior cingulate area is particularly active when novel combinations of actions are required of the individual, and it would appear that this area may store a sequence of actions, somewhat like the dorsolateral and ventrolateral prefrontal areas store location and shape attributes of an object. The temporary (working) memory component is presumed to be coupled to a retrieval mechanism that can enable the stored information to activate posterior cortical areas that express attention to object shape and location, and to activate anterior and posterior cortical areas that express attention to the execution of actions.

It may be conjectured that the top-down controls of attentional expressions can produce (through thalamocortical circuits) stronger enhancement effects than are elicited by sustained flow of information arising solely from sensory inputs. For example, two lights of changing luminosity placed on either side of a fixation point will each produce an enhancement effect in the information flowing upward through the visual cortical areas to location maps in the posterior parietal area. Instructions to attend to one of the lights result in projected activity from a prefrontal cortex map to a posterior parietal map to enhance the activity at the location of the attended light relative to the activity at the location of the non-attended light. Even if the attended light is less distinct (e.g., is smaller in size or shows only a slight oscillation of luminosity) than the unattended light, the additional activity arising from anterior prefrontal areas must enhance the site of the attended light over the activity at the site of the unattended light. If, however, the activity at the site of the more salient light remains dominant, then by definition attention is not being expressed at the site of the less salient light. Clearly, open questions remain as to exactly what posterior cortical areas will register the selective enhancements of attention projected from anterior cortical areas, what anterior cortical areas project this control, and what determines the level of attentional concentration in anterior cortical areas that induces corresponding levels of attentional concentration in posterior cortical areas.

In view of the fact that much, if not most, of the control of attention arises from voluntary processing, some readers may be tempted to attribute the "site of attention" to the prefrontal areas of brain, where converging evidence has indicated executive-like activities are processed. However, this book suggests that it is more useful to attribute the "site of attention" to the region or regions in the brain in which attention is being expressed and has its immediate effects, and to regard prefrontal brain sites as controlling (or regulating) attention in the sites where it is expressed. In a similar way, the "site of breathing" is attributed to the air passages, where breathing is expressed as the movement of air, rather than to the brainstem nuclei, where control of the breathing muscles resides. Although most examples of attentional expression in this book have been cases of perceptual processing within the posterior cortex, no reason has been given to restrict it to this area. Attention to actions, external and internal (e.g., shaping the forms of motor plans and shaping the forms of ideas), is assumed to be expressed in cortical pathways of the anterior cortex by the same general classes of algorithms (producing a difference in activity between the attended site and the non-attended sites) as the perceptual judgments of sensory stimuli are expressed in the posterior cortex.

A central theme of this book is that attention is expressed by the relative enhancement of information flow in particular pathways relative to the flow in surrounding pathways. A structure of the brain that produces this modulation of information flow is conjectured to be the thalamocortical circuit, which is found at virtually every area of the cerebral cortex. During the moments when the activity in a

cortical area that specializes in a perception, idea, or plan of action
is elevated sufficiently by the thalamocortical circuit, that perception, idea, or plan appears to "fill the mind." Being "mindful" of
a particular process within the brain is not just a spectator activity:
mindfulness, or attentional processing, shapes the way that mental
activity proceeds. Research in the cognitive neurosciences has only
begun to study the ways that thalamocortical enhancements of attention interact with the wide variety of information processing carried
out in the many areas of the cerebral cortex.

# Bibliography

Abramson, B. P., and L. M. Chalupa. 1988. Multiple pathways from the superior colliculus to the extrageniculate visual thalamus of the cat. *Journal of Comparative Neurology*, 271: 397–418.

Alexander, G. E., and M. D. Crutcher. 1990. Functional architecture of basal ganglia circuits: neural substrates of parallel processing. *Trends in Neurosciences*, 13: 266–271.

Allman, J. M., and J. H. Kaas. 1971. A representation of the visual field in the caudal third of the middle temporal gyrus of the owl monkey. *Brain Research*, 31: 85–105.

Allman, J. M., and J. H. Kaas. 1974. The organization of the second visual area (VII) in the owl monkey: A second-order transformation of the visual hemifield. *Brain Research*, 76: 247–265.

———— 1975. The dorsomedial cortical visual area: A third tier area in the occipital lobe of the owl monkey *(Aotus trivirgatus)*. *Brain Research*, 100: 473–487.

———— 1976. Representation of the visual field on the medial wall of the occipital-parietal cortex in the owl monkey. *Science*, 191: 572–575.

Allman, J. M, J. H. Kaas, R. H. Lane, and F. M. Miezin. 1972. A representation of the visual field in the inferior nucleus of the pulvinar in the owl monkey *(Aotus trivirgatus)*. *Brain Research*, 40: 291–302.

Allport, A. 1989. Visual attention. In *The Foundations of Cognitive Science*, ed. M. I. Posner, 631–682. Cambridge, MA: MIT Press.

Andersen, R. A. 1987. Inferior parietal lobule function in spatial perception and visuomotor integration. In *Handbook of Physiology*, sec. 1, *The Nervous System*, vol. 5: *Higher Functions of the Brain, Part 2*, ed. F. Plum, 483–518. Bethesda, MD: American Physiological Society.

———— 1989. Visual and eye movement functions of the posterior parietal cortex. *Annual Reviews of Neuroscience*, 12: 377–403.

Andersen, R. A., C. Asanuma, G. Essick, and R. M. Siegel. 1990. Corticocortical connections of anatomically and physiologically defined subdivisions within the inferior parietal lobule. *Journal of Comparative Neurology*, 296: 65–113.

Andersen, R. A., R. M. Bracewell, S. Barash, J. W. Gnadt, and L. Fogassi. 1990. Eye position effects on visual, memory, and sacccade-related activity in areas LIP and 7A of macaque. *Journal of Neuroscience*, 10: 1176–1196.

Andersen, R. A., G. K. Essick, and R. M. Siegel. 1985. Encoding of spatial location by posterior parietal neurons. *Science*, 230: 456–458.

Andersen, R. A., and D. Zipser. 1988. The role of the posterior parietal cortex in coordinate transformations for visuo-motor integration. *Canadian Journal of Physiology and Pharmacology*, 66: 488–501.

Araki, M., P. L. McGeer, and E. G. McGeer. 1984. Presumptive gamma-aminobutyric acid pathways from the midbrain to the superior colliculus studied by a combined horseradish-peroxidase-gamma-aminobutyric acid transaminase pharmacohistochemical method. *Neuroscience*, 13: 433–439.

Asanuma, C., R. A. Andersen, and W. M. Cowan. 1985. The thalamic relations of the caudal inferior parietal lobule and lateral prefrontal cortex in monkeys: Divergent cortical projections from cell clusters in the medial pulvinar nucleus. *Journal of Comparative Neurology*, 241: 357–381.

Bachinski, H. S., and V. R. Bachrach. 1980. Enhancement of perceptual sensitivity as the result of selectively attending to spatial locations. *Perception and Psychophysics*, 28: 241–248.

Bacon, W. F., and H. E. Egeth. 1994. Overriding stimulus-driven attentional capture. *Perception and Psychophysics*, 55: 485–496.

Baddeley, A. 1986. *Working Memory*. New York: Oxford University Press.

Baddeley, A. 1992. Working memory. *Science*, 255: 556–559.

Baizer, J. S., L. G. Ungerleider, and R. Desimone. 1991. *Society for Neuroscience Abstracts*, 332: 12.

Ballard, D. H. 1986. Cortical connections and parallel processing: Structure and function. *Behavioral and Brain Sciences*, 9: 67–120.

Ballard, D. H., G. E. Hinton, and T. J. Sejnowski. 1983. Parallel visual computation. *Nature*, 306: 21–26.

Barlow, H. B. 1972. Single units and sensation: A neuron doctrine for perceptual psychology? *Perception*, 1: 371–394.

Bates, J. F., F. A. W. Wilson, S. P. O'Scalaidhe, and P. S. Goldman-Rakic. 1994. Area TE connections with inferior prefrontal regions responsive to complex objects and faces. *Society for Neuroscience Abstracts*, 20: 1054.

Baylis, G. C., and J. Driver. 1992. Visual parsing and response competition: The effect of grouping factors. *Perception and Psychophysics*, 51: 145–162.

Baylis, G. C., E. T. Rolls, and C. M. Leonard. 1985. Selectivity between faces in the responses of a population of neurons in the cortex in the superior temporal sulcus of the monkey. *Brain Research,* 342: 91–102.

Behan, M., and L. P. Appell. 1992. Intrinsic circuitry in the cat superior colliculus: Projections from the superficial layers. *Journal of Comparative Neurology,* 315: 230–243.

Bender, D. B. 1981. Retinotopic organization of macaque pulvinar. *Journal of Neurophysiology,* 46: 672–693.

Benevento, L. A., and J. H. Fallon. 1975. The ascending projections of the superior colliculus in the rhesus monkey *(Macaca mulatta). Journal of Comparative Neurology,* 160: 339–362.

Benevento, L. A., and Rezak. 1976. The cortical projections of the inferior pulvinar and adjacent lateral pulvinar in the rhesus monkey: An autoradiographic study. *Brain Research,* 108: 1–24.

Bertelson, P., and R. Tisseyre. 1966. Choice reaction time as a function of stimulus versus response relative frequency of occurrence. *Nature,* 212: 1069–1070.

Blatt, G. J., R. A. Andersen, and G. R. Stoner. 1990. Visual receptive field organization and cortico-cortical connections of the lateral intraparietal area (Area LIP) in the macaque. *Journal of Comparative Neurology,* 299: 421–445.

Boch, R. A., and M. E. Goldberg. 1989. Participation of prefrontal neurons in the preparation of visually guided eye movements in the rhesus monkey. *Journal of Neurophysiology,* 61: 1064–1084.

Bonnel, A. M., C.-A. Possamai, and M. Schmitt. 1987. Early modulation of visual input: A study of attentional strategies. *Quarterly Journal of Experimental Psychology,* 39: 757–776.

Boschert, J. R. F. Hink, and L. Deecke. 1983. Finger movement– versus toe movement–related potentials: Further evidence for supplementary motor area (SMA) participation prior to voluntary action. *Experimental Brain Research,* 52: 73–80.

Boussaoud, D., L. G. Ungerleider, and R. Desimone. 1990. Pathways for motion analysis: Cortical connections of the medial superior temporal and fundus of the superior temporal visual areas in the macaque. *Journal of Comparative Neurology,* 296: 462–495.

Briand, K. A., and R. M. Klein. 1987. Is Posner's "beam" the same as Treisman's "glue"?: On the relation between visual orienting and feature integration theory. *Journal of Experimental Psychology: Human Perception and Performance,* 13: 228–241.

Broadbent, D. A. 1958. *Perception and Communication.* London: Pergamon Press.

Broadbent, D. E. 1971. *Decision and Stress.* New York: Academic Press.

Broadbent, D. E., and M. Gregory. 1964. Accuracy of recognition for speech presented to the right and left ears. *Quarterly Journal of Experimental Psychology,* 16: 359–360.

Bruce, C. J., R. Desimone, and C. G. Gross. 1981. Visual properties of neurons in a polysensory area in superior temporal sulcus of the macaque. *Journal of Neurophysiology,* 46: 369–384.

Bruce, C. J., and M. E. Goldberg. 1985. Prefrontal eye fields. I. Single neurons discharging before saccades. *Journal of Neurophysiology,* 53: 603–635.

Brunia, C. H. M. 1993. Waiting in readiness: Gating in attention and motor preparation. *Psychophysiology,* 30: 327–339.

Bundesen, C. 1990. A theory of visual attention. *Psychological Review,* 97: 523–547.

Bundesen, C., L. F. Pedersen, and A. Larsen. 1984. Measuring efficiency of selection from briefly exposed visual displays: A model for partial report. *Journal of Experimental Psychology: Human Perception and Performance,* 10: 329–339.

Burton, H., and E. G. Jones. 1976. The posterior thalamic region and its cortical projection in New World and Old World monkeys. *Journal of Comparative Neurology,* 168: 249–301.

Bush, R. R., and F. Mosteller. 1951. A mathematical model for simple learning. *Psychological Review,* 58: 313–323.

Bushnell, M. C., M. E. Goldberg, and D. L. Robinson. 1981. Behavioral enhancement of visual responses in monkey cerebral cortex. I. Modulation in posterior parietal cortex related to selective attention. *Journal of Neurophysiology,* 46: 755–772.

Carlsen, J. C. 1981. Some factors which influence melodic expectancy. *Psychomusicology,* 1: 12–29.

Casagrande, V. A., J. K. Harting, W. C. Hall, I. T. Diamond, and G. F. Martin. 1972. Superior colliculus of the tree shrew: A structural and functional subdivision into superficial and deep layers. *Science,* 177: 444–447.

Cattell, J. M. 1893. Aufmerksamkeit und Reaction. *Philosophische Studien,* 8: 403–406. Trans. in *James McKeen Cattell: Man of Science,* ed. A. Y. Poffenberger. New York: Arno Press, 1973.

Cavada, C., and P. S. Goldman-Rakic. 1989. Posterior parietal cortex in rhesus monkey: II. Evidence for segregated corticocortical networks linking sensory and limbic areas with the frontal lobe. *Journal of Comparative Neurology,* 287: 422–445.

Chagnac-Amitai, Y., H. J. Luhmann, and D. A. Prince. 1990. Burst generating and regular spiking layer 5 pyramidal neurons of rat neocortex have different morphological features. *Journal of Comparative Neurology,* 296: 598–613.

Chalupa, L. M. 1977. A review of cat and monkey studies implicating the pulvinar in visual function. *Behavioral Biology*, 20: 149–167.

—— 1991. Visual function of the pulvinar. In *The Neural Basis of Visual Function*, ed. A. G. Leventhal. Boca Raton: CRC Press.

Cheal, M., and D. Lyon. 1989. Attention effects on form discrimination at different eccentricities. *Quarterly Journal of Experimental Psychology*, 41A: 719–746.

Chelazzi, L., E. K. Miller, J. Duncan, and R. Desimone. 1993. A neural basis for visual search in inferior temporal cortex. *Nature*, 363: 345–347.

Chelazzi, L., E. K. Miller, A. Lueschow, and R. Desimone. 1993. Dual mechanisms of short-term memory: Ventral prefrontal cortex. *Society for Neuroscience Abstracts*, 19: 975.

Cherry, E. C. 1953. Some experiments on the recognition of speech with one and with two ears. *Journal of the Acoustical Society of America*, 25: 975–979.

Churchland, P. S. 1986. *Neurophilosophy*. Cambridge, MA: MIT Press.

Churchland, P. S., C. Koch, and T. J. Sejnowski. 1990. What is computational neuroscience? In *Computational Neuroscience*, ed. E. L. Schwartz, 46–55. Cambridge, MA: MIT Press.

Churchland, P. S., and T. J. Sejnowski. 1992. *The Computational Brain*. Cambridge, MA: MIT Press.

Cohen, J. D., K. Dunbar, and J. L. McClelland. 1990. On the control of automatic processes: A parallel distributed processing account of the Stroop effect. *Psychological Review*, 97: 332–361.

Colby, C. L. 1991. The neuroanatomy and neurophysiology of attention. *Journal of Child Neurology*, S90–S118.

Colby, C. L., J. R. Duhamel, and M. E. Goldberg. 1993. Ventral intraparietal area of the macaque: Anatomic location and visual response properties. *Journal of Neurophysiology*, 69: 902–914.

Conley, M., and D. Raczkowski. 1990. Sublaminar organization within layer VI of the striate cortex in *Galago*. *Journal of Comparative Neurology*, 302: 425–436.

Corbetta, M., F. M. Miezin, S. Dobmeyer, G. L. Shulman, and S. E. Petersen. 1991. Selective and divided attention during visual discrimination of shape, color, and speed: Functional anatomy by positron emission tomography. *Journal of Neuroscience*, 11: 2383–2402.

Corbetta, M., F. M. Miezin, G. L. Shulman, and S. E. Petersen. 1993. A PET study of visuospatial attention. *Journal of Neuroscience*, 13: 1202–1226.

Cowey, A., and D. G. Gross. 1970. Effects of foveal prestriate and inferotemporal lesions on visual discrimination by rhesus monkeys. *Experimental Brain Research*, 11: 128–144.

Craik, K. J. W. 1947. Theory of the human operator in control systems. I. The operator as an engineering system. *British Journal of Psychology,* 38: 56–61.

Crick, F. 1984. The function of the thalamic reticular complex: The search-light hypothesis. *Proceedings of the National Academy of Sciences (USA),* 81: 4586–4590.

Crick, F., and C. Koch. 1990. Towards a neurobiological theory of consciousness. *Seminars in the Neurosciences,* 2: 263–275.

Critchley, M. 1953. *The Parietal Lobes.* London: Edward Arnold.

Curcio, C. A., K. R. Sloan, Jr., O. Packer, A. E. Hendrickson, and R. E. Kalina. 1987. Distribution of cones in human and monkey retina: Individual variability and radial asymmetry. *Science,* 236: 579.

Damasio, A. 1985. Disorders of complex visual processing: Agnosias, achromotopsia, Baliant's syndrome, and related difficulties of orientation and construction. In *Principles of Behavioral Neurology,* ed. M. M. Mesulam, 259–288. Philadelphia: F. A. Davis.

Damasio, A. R., and A. L. Benton. 1979. Impairment of hand movements under visual guidance. *Neurology,* 29: 170–174.

Damasio, H., and A. Damasio. 1989. *Lesion Analysis in Neuropsychology.* Oxford: Oxford University Press.

Damasio, A. R., D. Tranel, and H. Damasio. 1992. Verbs but not nouns: Damage to left temporal cortices impairs access to nouns but not verbs. *Society for Neuroscience Abstracts,* 18: 387.

Desimone, R. 1991. Face-selective cells in the temporal cortex of monkeys. *Journal of Cognitive Neuroscience,* 3: 1–8.

——— 1992. Neural circuits for visual attention in the primate brain. In *Neural Networks for Vision and Image Processing,* ed. G. A. Carpenter and S. Grossberg, 343–364. Cambridge, MA: MIT Press.

Desimone, R., T. D. Albright, C. G. Gross, and C. Bruce. 1984. Stimulus selective properties of inferior temporal neurons in the macaque. *Journal of Neuroscience,* 4: 2051–2062.

Desimone, R., J. Fleming, and D. G. Gross. 1980. Prestriate afferents to inferior temporal cortex: An HRP study. *Brain Research,* 184: 41–55.

Desimone, R., and C. G. Gross. 1979. Visual areas in the temporal cortex of the macaque. *Brain Research,* 178: 363–380.

Desimone, R., L. Li, S. Lehky, L. G. Ungerlieder, and M. Mishkin. 1990. Effects of V4 lesions on visual discrimination performance and on responses of neurons in inferior temporal cortex. *Society for Neuroscience Abstracts,* 16: 621.

Desimone, R., and S. J. Schein. 1987. Visual properties of neurons in area V4 of the macaque: Sensitivity to stimulus form. *Journal of Neurophysiology,* 57: 835–868.

Desimone, R., S. J. Schein, J. Moran, and L. G. Ungerleider. 1985. Contour, color and shape analysis beyond the striate cortex. *Vision Research,* 25: 441–452.

Desimone, R., M. Wessinger, L. Thomas, and W. Schneider. 1989. Effects of deactivation of lateral pulvinar or superior colliculus on the ability to selectively attend to a visual stimulus. *Society of Neuroscience Abstracts,* 15: 162.

———— 1990. Attentional control of visual perception: Cortical and subcortical mechanisms. *Cold Spring Harbor Symposia on Quantitative Biology,* 55: 963–971.

Desimone, R., and L. G. Ungerleider. 1989. Neural mechanisms of visual processing in monkeys. In *Handbook of Neuropsychology,* vol. 2, ed. F. Boller and J. Grafman, 267–299. Amsterdam: Elsevier.

Deutsch, J. A., and D. Deutsch. 1963. Attention: Some theoretical considerations. *Psychological Review,* 70: 80–90.

DeYoe, E. A., J. Knierim, D. Sagi, B. Julesz, and D. Van Essen. 1986. Single unit responses to static and dynamic texture patterns in macaque V2 and V1 cortex. Association for Research in Vision and Ophthalmology Incorporated (Abstract).

Diamond, A., and P. S. Goldman-Rakic. 1989. Comparison of human infants and rhesus monkeys on Piaget's AB task: Evidence for dependence on dorsolateral prefrontal cortex. *Experimental Brain Research,* 74: 24–40.

Dick, A., A. Kaske, and O. D. Creutzfeldt. 1991. Topographical and topological organization of the thalamocortical projection to the striate and prestriate cortex in the marmoset. *Experimental Brain Research,* 84: 233–253.

Downing, C. 1988. Expectancy and visual-spatial attention: Effects on perceptual quality. *Journal of Experimental Psychology: Human Perception and Performance,* 14: 188–202.

Downing, C., and S. Pinker. 1985. The spatial structure of visual attention. In *Perception and Performance XI,* ed. M. I. Posner and O. S. M. Marin, 171–188. Hillsdale, NJ: Erlbaum.

Dubner, R., and S. M. Zeki. 1971. Response properties and receptive fields of cells in an anatomically defined region of the superior temporal sulcus in the monkey. *Brain Research,* 35: 528–532.

Duhamel, J., C. L., and M. E. Goldberg. 1992. The updating of the representation of visual space in parietal cortex by intended eye movements. *Science,* 255: 90–92.

Duncan, J. 1980. The locus of interference in the perception of simultaneous stimuli. *Psychological Review,* 87: 272–300.

———— 1984. Selective attention and the organization of visual information. *Journal of Experimental Psychology: General,* 113: 501–517.

—— 1994. Attention, intelligence, and the frontal lobes. In *The Cognitive Neurosciences*, ed. M. S. Gazzaniga. Cambridge, MA: MIT Press.

Duncan, J., and G. W. Humphreys. 1989. Visual search and stimulus similarity. *Psychological Review*, 96: 433–458.

Eason, R. G. 1981. Visual evoked potential correlates of early neural filtering during selective attention. *Bulletin of the Psychonomic Society*, 18: 203–206.

Egeth, H., R. Virzi, and H. Garbart. 1984. Searching for conjunctively defined targets. *Journal of Experimental Psychology: Human Perception and Performance*, 10: 32–39.

Egly, R., J. Driver, and R. D. Rafal. 1994. Shifting visual attention between objects and locations: Evidence from normal and parietal lesion subjects. *Journal of Experimental Psychology: General*, 123: 161–177.

Eimer, M. 1994. "Sensory gating" as a mechanism for visuospatial orienting: Electrophysiological evidence for trial-by-trial cuing experiments. *Perception and Psychophysics*, 55: 667–675.

Eriksen, B. A., and C. W. Eriksen. 1974. Effects of noise letters upon the identification of a target letter in a nonsearch task. *Perception and Psychophysics*, 16: 143–149.

Eriksen, C. W., and B. A. Eriksen. 1979. Target redundancy in visual search: Do repetitions of the target within the display impair processing? *Perception and Psychophysics*, 26: 195–205.

Eriksen, C. W., and J. E. Hoffman. 1972. Temporal and spatial characteristics of selective encoding from visual displays. *Perception and Psychophysics*, 12: 201–204.

—— 1973. The extent of processing of noise elements during selective coding from visual displays. *Perception and Psychophysics*, 14: 155–160.

Eriksen, C. W., and T. D. Murphy. 1987. Movement of attentional focus across the visual field: A critical look at the evidence. *Perception and Psychophysics*, 42: 299–305.

Eriksen, C. W., and J. D. St. James. 1986. Visual attention within and around the field of focal attention: A zoom lens model. *Perception and Psychophysics*, 40: 225–240.

Eriksen, C. W., and D. W. Schultz. 1979. Information processing in visual search: A continuous flow conception and experimental results. *Perception and Psychophysics*, 25: 249–263.

Eriksen, C. W., and J. Webb. 1989. Shifting of attentional focus within and about a visual display. *Perception and Psychophysics*, 42: 60–68.

Eriksen, C. W., and Y. Yeh. 1985. Allocation of attention in the visual field. *Journal of Experimental Psychology: Human Perception and Performance*, 11: 583–597.

Estes, W. K. 1950. Toward a statistical theory of learning. *Psychological Review,* 57: 94–107.

——— 1982. Similarity-related channel interactions in visual processing. *Journal of Experimental Psychology: Human Perception and Performance,* 8: 353–382.

Estes, W. K., and J. H. Straughan. 1954. Analysis of a verbal conditioning situation in terms of statistical learning theory. *Journal of Experimental Psychology,* 47: 225–234.

Falmagne, J. C., and J. Theios. 1969. On attention and memory in reaction time experiments. *Acta Psychologia,* 30: 316–323.

Farah, M. J. 1989. The neural basis of mental imagery. *Trends in Neurosciences,* 12: 395–399.

Farah, M. J., J. L. Brunn, A. B. Wong, M. A. Wallace, and P. A. Carpenter. 1990. Frames of reference for allocation of attention to space: Evidence from the neglect syndrome. *Neuropsychologia,* 28: 335–347.

Felleman, D. J., and D. C. Van Essen. 1991. Distributed hierarchical processing in the primate cerebral cortex. *Cerebral Cortex,* 1: 1–47.

Ferster, D., and S. Levay. 1978. The axonal arborization of lateral geniculate neurons in the striate cortex of the cat. *Journal of Comparative Neurology,* 182: 923–944.

Fischer, B., and R. Boch. 1981. Enhanced activation of neurons in prelunate cortex before visually guided saccades of trained rhesus monkeys. *Experimental Brain Research,* 44: 129–137.

Fischer, B., and H. Weber. 1993. The time of secondary saccades to primary targets. *Experimental Brain Research,* 97: 356–360.

Fodor, J. A. 1983. *The Modularity of Mind.* Cambridge, MA: MIT Press.

Foote, S. L., and J. H. Morrison. 1987. Extrathalamic modulation of cortical function. *Annual Review of Neuroscience,* 10: 67–95.

Fox, P. T., J. S. Perlmutter, and M. E. Raichle. 1985. A stereotactic method of anatomical localization for positron emission tomography. *Journal of Computer Assisted Tomography,* 9: 141–153.

Freeman, W. J. 1975. *Mass Action in the Nervous System.* New York: Academic Press.

Fries, W. 1984. Cortical projections to the superior colliculus in the macaque monkey: A retrograde study using horseradish peroxidase. *Journal of Comparative Neurology,* 230: 55–76.

Frith, C. D., K. Friston, P. F. Liddle, and R. S. J. Frackowiak. 1991. Willed action and the prefrontal cortex in man: A study with PET. *Proceedings of the Royal Society of London (Biology),* 244: 241–246.

Fruhstorfer, H., P. Saveri, and T. Jarvilehto. 1970. Short-term habituation of the auditory evoked response in man. *Electroencephalography and Clinical Neurophysiology,* 28: 153–161.

Fujita, I., K. Tanaka, M. Ito, and C. Kang. 1992. Columns for visual features of objects in monkey inferotemporal cortex. *Nature,* 360: 343–346.

Funahashi, S., C. J. Bruce, and P. S. Goldman-Rakic. 1989. Mnemonic coding of visual space in the monkey's dorsolateral prefrontal cortex. *Journal of Neurophysiology,* 61: 331–349.

———— 1990. Visuospatial coding in primate prefrontal neurons revealed by oculomotor paradigms. *Journal of Neurophysiology,* 63: 814–831.

———— 1991. Neuronal activity related to saccadic eye movements in the monkey's dorsolateral prefrontal cortex. *Journal of Neurophysiology,* 65: 1464–1483.

Gattass, R., and R. Desimone. 1991. Attention-related responses in the superior colliculus of the macaque. *Society for Neuroscience Abstracts,* 17: 545.

———— 1992. Stimulation of the superior colliculus (SC) shifts the focus of attention in the macaque. *Society for Neuroscience Abstracts,* 18: 703.

Gattass, R., C. G. Gross, and J. H. Sandell. 1981. Visual topography of V2 in the macaque. *Journal of Comparative Neurology,* 21: 5129–5139.

Gattass, R., E. Oswaldo-Cruz, and A. P. B. Sousa. 1978. Visuotopic organization of the *Cebus* pulvinar: A double representation of the contralateral hemifield. *Brain Research,* 152: 1–16.

Gattass, R, A. P. B. Sousa, and C. G. Gross. 1988. Visuotopic organization and extent of V3 and V4 of the macaque. *Journal of Neuroscience,* 8: 1831–1845.

Gawne, T. J., E. N. Eskandar, and B. J. Richmond. 1992. The heterogeneity of adjacent neurons in inferior temporal cortex. *Society for Neuroscience Abstracts,* 18: 147.

Gazzaniga, M. S. 1970. *The Bisected Brain.* New York: Appleton.

———— 1985. *The Social Brain: Discovering the Networks of the Mind.* New York: Basic Books.

Giguere, M., and P. S. Goldman-Rakic. 1988. Mediodorsal nucleus: Areal, laminar, and tangential distribution of afferents and efferents in the frontal lobe of rhesus monkeys. *Journal of Comparative Neurology,* 277: 195–213.

Gilbert, C. D. 1983. Microcircuitry of the visual cortex. *Annual Reviews of Neuroscience,* 6: 217–247.

Gilbert, C. D., and J. P. Kelly. 1975. The projections of cells in different layers of the cat's visual cortex. *Journal Comparative Neurology,* 163: 81–105.

Gilbert, C. D., and T. N. Wiesel. 1983. Clustered intrinsic connections in cat visual cortex. *Journal of Neuroscience,* 3: 1116–1133.

Gnadt, J. W., and R. A. Andersen. 1988. Memory related motor planning activity in posterior parietal cortex of macaque. *Experimental Brain Research,* 70: 216–220.

Goldberg, M. E., and M. C. Bushnell. 1981. Behavioral enhancement of visual responses in monkey cerebral cortex. II. Modulation in frontal eye fields specifically related to saccades. *Journal of Neurophysiology* 46: 773.

Goldberg, M. E., and C. L. Colby. 1989. The neurophysiology of spatial vision. In *Handbook of Neuropsychology*, vol. 2, ed. F. Boller and J. Grafman, 301–315. Amsterdam: Elsevier.

—— 1992. Oculomotor control and spatial processing. *Current Opinion in Neurobiology*, 2: 198–202.

Goldberg, M. E., C. L. Colby, and J.-R. Duhamel. 1990. Representation of visuomotor space in the parietal lobe of the monkey. *Cold Spring Harbor Symposia on Quantitative Biology*, 55: 729–739.

Goldberg, M. E., and R. H. Wurtz. 1970. Effects of eye movement and stimulus on units in monkey superior colliculus. *Federation Proceedings*, 29: 453.

—— 1972. Activity of superior colliculus in behaving monkey. II. Effect of attention on neuronal responses. *Journal of Neurophysiology*, 35: 560–574.

Goldman, P. S., and W. J. H. Nauta. 1976. Autoradiographic demonstration of a projection from prefrontal association cortex to the superior colliculus in the rhesus monkey. *Brain Research*, 116: 145–149.

Goldman-Rakic, P. S. 1987. Circuitry of primate prefrontal cortex and regulation of behavior by representational memory. In *Handbook of Physiology*, sec. 1, *The Nervous System*, vol. 5: *Higher Functions of the Brain, Part 2*, ed. F. Plum, 373–417. Bethesda, MD: American Physiological Society.

—— 1988. Topography of cognition: Parallel distributed networks in primate association cortex. *Annual Review of Neuroscience*, 11: 137–156.

Goldman-Rakic, P. S., M. Chafee, and H. Friedman. 1993. Allocation of function in distributed circuits. In *Brain Mechanisms of Perception and Memory: From Neuron to Behavior*, ed. T. Ono, L. R. Squire, M. E. Raichle, D. I. Perrett, and M. Fukuda, 445–456. New York: Oxford University Press.

Goldman-Rakic, P. S., and L. J. Porrino. 1985. The primate mediodorsal (MD) nucleus and its projections to the frontal lobe. *Journal of Comparative Neurology*, 242: 535–560.

Goodman, S. J., and R. A. Andersen. 1989. Microstimulation of a neural-network model for visually guided saccades. *Journal of Cognitive Neuroscience*, 4: 317–326.

Grace, A. A., and B. S. Bunney. 1979. Paradoxical GABA excitation of nigral dopaminergic cells: Indirect mediation through reticulata inhibitory neurons. *European Journal of Pharmacology*, 59: 211–218.

Graham, J., C. S. Lin, and J. H. Kaas. 1979. Subcortical projections of six visual cortical areas in the owl monkey, *Aotus trivirgatus*. *Journal of Comparative Neurology*, 187: 557–580.

Graybiel, A. M., and C. W. Ragsdale Jr. 1979. Fiber connections of the basal ganglia. In *Development and Chemical Specificity of Neurons*, ed. M. Cuenod, G. W. Kreutzberg, and F. E. Bloom, 239–283. Amsterdam: Elsevier.

Green, D. M., and R. D. Luce. 1974. Timing and counting mechanisms in auditory discrimination and reaction time. In *Contemporary Developments in Mathematical Psychology*, vol. 2, ed. D. H. Krantz, R. C. Atkinson, R. D. Luce, and P. Suppes, 372–415. San Francisco: Freeman.

Green, D. M., and J. A. Swets. 1966. *Signal Detection Theory and Psychophysics*. New York: Wiley.

Gross, C. G., C. J. Bruce, R. Desimone, J. Fleming, and R. Gattass. 1981. Cortical visual areas of the temporal lobe. In *Cortical Sensory Organization*, vol. 2, ed. C. N. Woolsey, 187–216. Englewood Cliffs, NJ: Humana Press.

Gross, C. G., C. E. Rocha-Miranda, and D. B. Bender. 1972. Visual properties of neurons in inferotemporal cortex of the macaque. *Journal of Neurophysiology*, 35: 96–111.

Guthrie, E. R. 1959. Association by contiguity. In *Psychology: A Study of a Science*, vol. 2, ed. S. Koch, 158–195. New York: McGraw-Hill.

Hackley, S. A., R. Schaffer, and J. Miller. 1990. Preparation for Donders' type Band C reaction tasks. *Acta Psychologica*, 74: 15–33.

Haenny, P. E., J. H. R. Maunsell, and P. H. Schiller. 1988. State dependent activity in monkey visual cortex: II. Retinal and extraretinal factors in V4. *Experimental Brain Research*, 69: 245–259.

Haenny, P. E., and P. H. Schiller. 1988. State dependent activity in monkey visual cortex: I. Single cell activity in V1 and V4 on visual tasks. *Experimental Brain Research*, 69: 225–244.

Harting, J. K., M. F. Huerta, A. J. Frankfurter, N. L. Strominger, and G. J. Royce. 1980. Ascending pathways from the monkey superior colliculus: An autoradiographic analysis. *Journal of Comparative Neurology*, 192: 853–882.

Hawkins, H. L, S. A. Hillyard, S. J. Luck, M. Mouloua, C. J. Downing, and D. P. Woodward. 1990. Visual attention modulates signal detectability. *Journal of Experimental Psychology: Human Perception and Performance*, 16: 802–811.

Haxby, J. V., C. L. Grady, B. Horwitz, L. G. Ungerleider, M. Mishkin, R. E. Carson, P. Herscovitch, M. B. Schapiro, and S. I. Rapoport. 1991. Dissociation of object and spatial visual processing pathways in human extrastriate cortex. *Proceedings of the National Academy of Sciences (USA)*, 88: 1621–1625.

Healy, A., S. Kosslyn, and R. M. Shiffrin. 1992. *From Learning Theory to Connectionist Theory.* Hillsdale, NJ: Erlbaum.

Hebb, D. O. 1949. *The Organization of Behavior.* New York: Wiley.

Heil, M., F. Rosler, and E. Henninghausen. 1993. Imagery-perception interaction depends on the shape of the image: A reply to Farah (1989). *Journal of Experimental Psychology: Human Perception and Performance,* 19: 1313–1320.

Heilman, K. M., D. Bowers, H. B. Coslett, H. Whelan, and R. T. Watson. 1985. Directional hypokinesia: Prolonged reaction times for leftward movements in patients with right hemisphere lesions and neglect. *Neurology,* 35: 855–859.

Heilman, K. M., R. T. Watson, E. Valenstein, and M. E. Goldberg. 1987. Attention: Behavior and neural mechanisms. In *The Handbook of Physiology,* sec. 1, *The Nervous System,* vol. 5: *Higher Functions of the Brain, Part 2,* ed. F. Plum. Bethesda, MD: American Physiological Society.

Heinze, H.-J., S. J. Luck, T. F. Münte, A. Gös, G. R. Mangun, and S. A. Hillyard. 1994. Attention to adjacent and separate positions in space: An electrophysiological analysis. *Perception and Psychophysics,* 56: 42–52.

Henderson, J. M. 1991. Stimulus discrimination following covert attentional orienting to an exogenous cue. *Journal of Experimental Psychology: Human Perception and Performance,* 17: 91–106.

Henderson, J. M., A. Pollatsek, and K. Rayner. 1989. Covert visual attention and extrafoveal information use during object identification. *Perception and Psychophysics,* 45: 196–208.

Hernandez-Peon, R. 1960. Neurophysiological correlates of habituation and other manifestations of plastic inhibition (internal inhibition). In *The Moscow Colloquium on Electroencephalography of Higher Nervous Activity,* ed. H. H. Jasper and G. D. Smirnov. *Electroenchephalography and Clinical Neurophysiology,* Supplement No. 13, 101–114.

———— 1966. Physiological mechanisms in attention. In *Frontiers in Physiological Psychology,* ed. R. W. Russell, 121–147. New York: Academic Press.

Heywood, C. A., and A. Cowey. 1987. On the role of cortical area V4 in the discrimination of hue and pattern in macaque monkeys. *Journal of Neuroscience,* 7: 2601–2617.

Hicks, T. P., S. Molotchnikoff, and T. Ono. 1993. *The Visually Responsive Neuron: From Basic Neurophysiology to Behavior.* New York: Elsevier.

Hikosaka, O., and R. H. Wurtz. 1983. Visual and oculomotor functions of monkey substantia nigra pars reticulata IV: Relation of substantia nigra to superior colliculus. *Journal of Neurophysiology,* 49: 1285–1301.

———— 1985. Modification of saccadic eye movements by GABA-related substances. I. Effect of muscimol and bicuculline in monkey superior colliculus. *Journal of Neurophysiology,* 53: 266–291.

————— 1989. The basal ganglia. In *The Neurobiology of Saccadic Eye Movements,* ed. R. H. Wurtz and M. E. Goldberg. New York: Elsevier.

Hillyard, S. A., and M. Kutas. 1983. Electrophysiology of cognitive processing. *Annual Review of Psychology,* 34: 33–61.

Hillyard, S. A., and G. R. Mangun. 1987. Sensory gating as a physiological mechanism for visual selective attention. In *Current Trends in Event-related Potential Research,* ed. R. Johnson, Jr., R. Parasuraman, and J. W. Rohrbaugh, 61–67. New York: Elsevier.

————— 1994. Neural systems mediating selective attention. In *The Cognitive Neurosciences,* ed. M. Gazzaniga. Cambridge, MA: MIT Press.

Hillyard, S. A., and T. F. Münte. 1984. Selective attention to color and location: An analysis with event-related brain potentials. *Perception and Psychophysics,* 36: 185–198.

Hinton, G. E., J. L. McClellan, and D. E. Rumelhart. 1986. Distributed representations. In *Parallel Distributed Processing: Explorations in the Microstructure of Cognition,* vol. 1.: *Foundations,* ed. D. E. Rumelhart and J. L. McClelland. Cambridge, MA: MIT Press.

Hirai, T., and E. G. Jones. 1989. A new parcellation of the human thalamus on the basis of histochemical staining. *Brain Research Reviews,* 14: 1–34.

Hobson, J. A., and M. Steriade. 1986. The neuronal basis of behavioral state control. In *Handbook of Physiology,* sec. 1, *The Nervous System,* vol. 4: *Intrinsic Regulatory Systems of the Brain,* ed. F. E. Bloom, 701–823. Bethesda, MD: American Physiological Society.

Hodgkin, A. L., and A. F. Huxley. 1952. A quantitative description of membrane current and its application to conduction and excitation in nerve. *Journal of Physiology* (London), 117: 500–544.

Holtzman, J. D., B. T. Volpe, and M. S. Gazzaniga. 1984. Spatial orientation following commissural section. In *Varieties of Attention,* ed. R. Parasuraman and D. R. Davies, 375–394. London: Academic Press.

Hubel, D. H., and M. S. Livingston. 1987. Segregation of form, color, and stereopsis in primate area 18. *Journal of Neuroscience,* 7: 3378–3415.

Huerta, M. F., and J. K. Harting. 1984a. Connectional organization of the superior colliculus. *Trends in Neurosciences,* 7: 286–289.

————— 1984b. The mammalian superior colliculus: Studies of its morphology and connections. In *The Comparative Neurology of the Optic Tectum,* ed. H. Vanegas, 687–773. New York: Plenum Press.

Hughes, H. C., and L. D. Zimba. 1987. Natural boundaries of the spatial spread of directed visual attention. *Neuropsychologia,* 25: 5–18.

Hyvarinen, J. 1982. Parietal association cortex: Posterior parietal lobe of the primate brain. *Physiological Reviews,* 62: 1060–1129.

Illing, R. B., and A. M. Graybiel. 1985. Convergence of afferents from frontal cortex and substantia nigra onto acetylcholinesterase-rich patches of the cat's superior colliculus. *Neuroscience,* 14: 455–482.

Intons-Petersen, M., and B. B. Roskos-Ewoldson. 1989. Sensory-perceptual qualities of images. *Journal of Experimental Psychology: Learning, Memory, and Cognition,* 15: 188–199.

Iwai, E. 1985. Neurophysiological basis of pattern vision in macaque monkeys. *Vision Research,* 25: 425–439.

James, W. 1890. *Principles of Psychology,* 2 vols. New York: Holt.

Johnson, R. R., and A. Burkhalter. 1994. Different microcircuits for forward corticocortical and LP (pulvinar) projections in rat extrastriate visual cortex. *Society for Neuroscience Abstracts,* 20: 427.

Jones, E. G. 1985. *The Thalamus.* New York: Plenum Press.

———— 1988. What are the local circuits? In *Neurobiology of Neocortex,* ed. P. Rakic and W. Singer, 137–152. Chichester: Wiley.

Jonides, J. 1980. Towards a model of the mind's eye's movement. *Canadian Journal of Psychology,* 34: 103–112.

———— 1981. Voluntary vs. automatic control over the mind's eye movement. In *Attention and Performance XI,* ed. M. I. Posner and O. S. M. Marin. Hillsdale, NJ: Erlbaum.

———— 1983. Further toward a model of the mind's eye's movement. *Bulletin of the Psychonomic Society,* 21: 247–250.

Jonides, J., E. E. Smith, R. A. Koeppe, E. Awh, S. Minoshima, and M. A. Mintun. 1993. Spatial working memory in humans as revealed by PET. *Nature,* 363: 623–625.

Jonides, J., and S. Yantis. 1988. Uniqueness of abrupt visual onset in capturing attention. *Perception and Psychophysics,* 43: 346–354.

Julesz, B. 1984. Toward an axiomatic theory of preattentive vision. In *Dynamic Aspects of Neocortical Function,* ed. G. M. Edelman, W. E. Gall, and W. M. Cowan, 586–610. New York: Wiley.

Julesz, B., and J. R. Bergen. 1983. Textons, the fundamental elements in preattentive vision and perception of textures. *Bell Systems Technical Journal,* 62: 1619–1645.

Juola, J. F., D. G. Bouwhuis, E. E. Cooper, and C. B. Warner. 1991. Control of attention around the fovea. *Journal of Experimental Psychology: Human Perception and Performance,* 17: 125–141.

Kahneman, D. 1973. *Attention and Effort.* Englewood Cliffs, NJ: Prentice-Hall.

Kahneman, D., and D. Chajczyk. 1983. Tests of the automaticity of reading: Dilution of Stroop effects by color-irrelevant stimuli. *Journal of Experimental Psychology: Human Perception and Performance,* 9: 497–509.

Kahneman, D., and A. Henik. 1977. Effects of visual grouping on immedi-

ate recall and selective attention. In *Attention and Performance VI*, ed.
S. Dornic, 307–332. Hillsdale, NJ: Erlbaum.

Kahneman, D., A. Treisman, and J. Burkell. 1983. The cost of visual filtering. *Journal of Experimental Psychology: Human Perception and Performance*,
9: 510–522.

Kahneman, D., A. Treisman, and B. J. Gibbs. 1992. The reviewing of object
files: Object-specific integration of information. *Cognitive Psychology*, 24:
179–219.

Karabelas, A. B., and A. K. Moschovakis. 1985. Nigral inhibitory termination
on efferent neurons of the superior colliculus: An intracellular horseradish peroxidase study in the cat. *Journal of Comparative Neurology*, 239:
309–329.

Kaske, A., A. Dick, and O. D. Creutzfeldt. 1991. The local domain for diver-
gence of subcortical afferents to the striate and extrastriate visual cortex in the common marmoset: A multiple labelling study. *Experimental
Brain Research*, 84: 254–265.

Keele, S. W., and W. T. Neill. 1978. Mechanisms of attention. In *Handbook
of perception*, vol. 9, ed. E. C. Carterette and M. P. Friedman, 3–47. New
York: Academic Press.

Kertzman, C., and D. L. Robinson. 1988. Contributions of the superior colliculus of the monkey to visual spatial attention. *Society of Neuroscience
Abstracts*, 14: 831.

Kikuchi, R., and E. Iwai. 1980. The locus of the posterior subdivision of the
inferotemporal visual learning area in the monkey. *Brain Research*, 198:
347–360.

Kinchla, R. A. 1977. The role of structural redundancy in the detection of
visual targets. *Perception and Psychophysics*, 22: 19–30.

King, A. J., and A. R. Palmer. 1985. Integration of visual and auditory information in bimodal neurons in the guinea-pig superior colliculus. *Experimental Brain Research*, 60: 492–500.

Kingstone, A. 1992. Combining expectancies. *Quarterly Journal of Experimental Psychology*, 44A: 69–104.

Kingstone, A., and R. M. Klein. 1993. On the relationship between overt
and covert orienting: The effect of attended and unattended offsets
on saccadic latencies. *Journal of Experimental Psychology: Human Perception and Performance*, 19: 1251–1265.

Kinsbourne, M. 1987. Mechanisms of unilateral neglect. In *Neurophysiological and Neuropsychological Aspects of Spatial Neglect*, ed. M. Jeannerod,
69–86. Amsterdam: Elsevier.

Klein, R. 1980. Does oculomotor readiness mediate cognitive control of
visual attention? In *Attention and Performance VIII*, ed. R. S. Nickerson,
259–276. Hillsdale, NJ: Erlbaum.

Klein, R. M., and E. Hansen. 1990. Chronometric analysis of spotlight failure in endogenous visual orienting. *Journal of Experimental Psychology: Human Perception and Performance,* 16: 790–801.

Klein, R. M., and A. Pontefract. 1994. Does oculomotor readiness mediate cognitive control of visual attention? Revisited. In *Attention and Performance XV: Conscious and Unconscious Processing,* ed. C. Umilta and M. Moscovitch. Cambridge, MA: MIT Press.

Koch, C., and S. Ullman. 1985. Shifts in selective visual attention: Towards the underlying neural circuitry. *Human Neurobiology,* 4: 219–227.

Koopman, B. O. 1957. Part III. The optimum distribution of searching effort. *Operations Research,* 5: 613–626.

Kornblum, S., and J. Requin. 1984. *Preparatory States and Processes.* Hillsdale, NJ: Erlbaum.

Koshino, H., C. B. Warner, and J. F. Juola. 1992. Relative effectiveness of central, peripheral, and abrupt-onset cues in visual attention. *Quarterly Journal of Experimental Psychology,* 45A: 609–631.

Kosslyn, S. M. 1980. *Image and Mind.* Cambridge: Harvard University Press.

——— 1988. Aspects of a cognitive neuroscience of mental imagery. *Science,* 240: 1621–1626.

——— 1994. *Image and Brain: The Resolution of the Imagery Debate.* Cambridge, MA: MIT Press.

Kosslyn, S. M., N. M. Alpert, W. L. Thompson, and V. Maljkovic. 1993. Visual mental imagery activates topographically organized visual cortex: PET investigations. *Journal of Cognitive Neuroscience,* 5: 263–287.

Kosslyn, S. M., and A. L. Sussman. 1994. Roles of imagery in perception: Or, there is no such thing as immaculate perception. In *The Cognitive Neurosciences,* ed. G. Gazzaniga. Cambridge, MA: MIT Press.

Kramer, A. F., and A. Jacobson. 1991. Perceptual organization and focused attention: The role of objects and proximity in visual processing. *Perception and Psychophysics,* 50: 267–284.

LaBerge, D. 1973. Identification of the time to switch attention: A test of a serial and a parallel model of attention. In *Attention and Performance IV,* ed. S. Kornblum. New York: Academic Press.

———. 1983. The spatial extent of attention to letters and words. *Journal of Experimental Psychology: Human Perception and Performance,* 9: 371–379.

——— 1990a. Attention. *Psychological Science,* 1: 156–162.

——— 1990b. Thalamic and cortical mechanisms of attention suggested by recent positron emission tomographic experiments. *Journal of Cognitive Neuroscience,* 2: 358–372.

——— 1994a. Computational and anatomical models of selective attention in object identification. In *The Cognitive Neurosciences,* ed. M. Gazzaniga. Cambridge, MA: MIT Press.

———— 1994b. Quantitative models of attention and response processes in shape identification tasks. *Journal of Mathematical Psychology,* 38: 198–243.

LaBerge, D., and V. Brown. 1989. Theory of attentional operations in shape identification. *Psychological Review,* 96: 101–124.

LaBerge, D., V. Brown, M. Carter, A. Hartley, and D. Bash. 1991. Reducing the effects of adjacent distractors by narrowing attention. *Journal of Experimental Psychology: Human Perception and Performance,* 17: 65–76.

LaBerge, D., and M. S. Buchsbaum. 1990. Positron emission tomographic measurements of pulvinar activity during an attention task. *Journal of Neuroscience,* 10: 613–619.

LaBerge, D., M. Carter, and V. Brown. 1992. A network simulation of thalamic circuit operations in selective attention. *Neural Computation,* 4: 318–331.

LaBerge, D., R. Legrand, and R. K. Hobbie. 1969. Functional identification of perceptual and response biases in choice reaction time. *Journal of Experimental Psychology,* 79: 295–299.

LaBerge, D., and S. J. Samuels. 1974. Toward a theory of automatic information processing in reading. *Cognitive Psychology,* 6: 293–323.

LaBerge, D., and J. R. Tweedy. 1961. Presentation probability and choice time. *Journal of Experimental Psychology,* 67: 71–79.

LaBerge, D., J. R. Tweedy, and J. Ricker. 1967. Selective attention: Incentive variables and choice time. *Psychonomic Science,* 8: 341–342.

LaBerge, D., P. Van Gelder, and J. I. Yellott. 1970. A cueing technique in choice reaction time. *Perception and Psychophysics,* 7: 57–62.

Laming, D. R. J. 1968. *Information Theory of Choice-Reaction Times.* London: Academic Press.

Lawrence, D. H. 1950. Acquired distinctiveness of cues: II. Selective association in a constant stimulus situation. *Journal of Experimental Psychology,* 40: 175–188.

Lehky, S. R., and T. J. Sejnowski. 1990. Neural model of stereoacuity and depth interpolation based on a distributed representation of stereo disparity. *Journal of Neuroscience,* 10: 2281–2299.

Leichnetz, G. R., R. F. Spencer, S. G. Hardy, and J. Astruc. 1981. The prefrontal corticotectal projection in the monkey: An anterograde and retrograde horseradish peroxidase study. *Neuroscience,* 6: 1023–1041.

Levay, S., and C. Gilbert. 1976. Laminar patterns of geniculocortical projection in the cat. *Brain Research,* 113: 1–19.

Liotti, M., P. T. Fox, and D. LaBerge. 1994. PET measurements of attention to closely spaced visual shapes. *Society for Neurosciences Abstracts,* 20: 354.

Llinas, R., and H. Jahnsen. 1982. Electrophysiology of mammalian thalamic neurons in vitro. *Nature,* 297: 406–408.

Livingston, M. S., and D. Hubel. 1984. Specificity of intrinsic connections in primate primary visual cortex. *Journal of Neuroscience,* 4: 2830–2835.

Luce, R. D. 1959. *Individual Choice Behavior: A Theoretical Analysis.* New York: Wiley.

———— 1977. Thurstone's discriminal processes fifty years later. *Psychometrika,* 42: 461–489.

———— 1986. *Response Times.* New York: Oxford University Press.

Luck, S. J, L. Chelazzi, S. A. Hillyard, and R. Desimone. 1992. Attentional modulation of responses in area V4 of the macaque. *Society for Neurosciences Abstracts,* 18: 147.

Lund, J. S., A. E. Hendrickson, M. P. Ogren, and E. A. Tobin. 1981. Anatomical organization of primate visual cortex area VII. *Journal of Comparative Neurology,* 202: 19–45.

Lynch, J. C., A. M. Graybiel, and L. J. Lobeck. 1985. The differential projection of two cytoarchitectonic subregions of the inferior parietal lobule of macaque upon the deep layers of the superior colliculus. *Journal of Comparative Neurology,* 235: 241–254.

Lynch, J. C., and J. W. McClaren. 1989. Deficits of visual attention and saccadic eye movements after lesions of parieto-occipital cortex in monkeys. *Journal of Neurophysiology,* 61: 74–90.

Lynch, J. C., V. B. Mountcastle, W. H. Talbot, and C. T. Yin. 1977. Parietal lobe mechanisms for directed visual attention. *Journal of Neurophysiology,* 40: 362–389.

Lynn, R. 1966. *Attention, Arousal, and the Orientation Reaction.* Oxford: Pergamon.

Mackworth, J. F. 1969. *Vigilance and Habituation.* Harmondsworth, Eng.: Penguin Books.

Mangun, G. R., J. C. Hansen, and S. A. Hillyard. 1987. The spatial orienting of visual attention: Sensory facilitation or response bias? In *Current Trends in Event-related Potential Research,* ed. R. Johnson, Jr., J. W. Rohrbaugh, and R. Parasuraman, 118–124. New York: Elsevier.

Mangun, G. R., and S.A. Hillyard, 1987. The spatial allocation of visual attention as indexed by event-related brain potentials. *Human Factors,* 29: 195–211.

Mangun, G. R., S. A. Hillyard, and S. J. Luck. 1992. Electrocortical substates of visual selective attention. In *Attention and Performance XIV,* ed. D. Meyer and S. Kornblum, 219–243. Hillsdale, NJ: Erlbaum.

Marr, D. 1982. *Vision.* San Francisco: W. H. Freeman.

Marr, D., and E. Hildreth. 1980. Theory of edge detection. *Proceedings of the Royal Society* (B), 207: 187–217.

McGinn, C. 1990. *The Problem of Consciousness.* Oxford: Blackwell.

McIlwain, J. T. 1991. Distributed spatial coding in the superior colliculus: A review. *Visual Neuroscience,* 6: 3–13.

Meadows, J. C. 1974. The anatomical basis of prosopagnosia. *Journal of Neurological Neurosurgical Psychiatry,* 37: 489–501.

Meredith, M. A., and B. E. Stein. 1985. Descending efferents from the superior colliculus relay integrated multisensory information. *Science,* 227: 657–659.

——— 1990. The visuotopic component of the multisensory map in the deep laminae of the cat superior colliculus. *Journal of Neuroscience,* 10: 3727–3742.

Mesulam, M. M. 1981. A cortical network for directed attention and unilateral neglect. *Annals of Neurology,* 10: 309–325.

Miller, J. 1989. The control of attention by abrupt visual onsets and offsets. *Perception and Psychophysics,* 45: 567–571.

Miller, J., and A.-M. Bonnel. 1994. Switching or sharing in dual-task line-length discrimination. *Perception and Psychophysics,* 56: 431–446.

Mishkin, M., L. G. Ungerleider, and K. A. Macko. 1983. Object vision and spatial vision: Two cortical pathways. *Trends in Neurosciences,* 6: 414–417.

Miyashita, Y., and H. S. Chang. 1988. Neuronal correlate of pictorial short-term memory in the primate temporal cortex. *Nature,* 331: 68–70.

Mohler, C., and R. H. Wurtz. 1976. Organization of monkey superior colliculus: Intermediate layer cells discharging before eye movements. *Journal of Neurophysiology,* 39: 722–744.

Montero, V. M., and W. Singer. 1985. Ultrastructural identification of somata and neural processes immunoreactive to antibodies against glutamic acid decarboxylase (GAD) in the dorsal lateral geniculate nucleus of the cat. *Experimental Brain Research,* 59: 151–165.

Mooney, R. D., X. Huang, and R. W. Rhoades. 1990. Effects of inactivation of the superficial laminae upon the visual responsivity of deep layer neurons in the hamster's superior colliculus. *Society for Neuroscience Abstracts,* 16: 109.

Mooney, R. D., M. M. Nikoletseas, P. R. Hess, Z. Allen, A. C. Lewin, and R. W. Rhodes. 1988. The projections from the superficial to the deep layers of the superior colliculus: An intracellular horseradish peroxidase injection study in the hamster. *Journal of Neuroscience,* 8: 1384.

Moran, J., and R. Desimone. 1985. Selective attention gates visual processing in the extrastriate cortex. *Science,* 229: 782–784.

Moray, N. 1959. Attention in dichotic listening: Affective cues and the influence of instructions. *Quarterly Journal of Experimental Psychology,* 11: 56–60.

Motter, B.C. 1993. Focal attention produces spatially selective processing in visual cortical areas V1, V2, and V4 in the presence of competing stimuli. *Journal of Neurophysiology,* 70: 909–919.

Motter, B. C., and V. B. Mountcastle. 1981. The functional properties of the light-sensitive neurons of the posterior parietal cortex studied in waking monkeys: Foveal sparing and opponent vector organization. *Journal of Neuroscience,* 1: 3–26.

Mountcastle, V. B. 1978. Brain mechanisms of directed attention. *Journal of the Royal Society of Medicine,* 71: 14–27.

Mountcastle, V. B., J. C. Lynch, A. Georgopoulos, H. Sakata, and C. Acuna. 1975. Posterior parietal association cortex of the monkey: Command functions for operations within extrapersonal space. *Journal of Neurophysiology,* 38: 871–907.

Müller, H. H., and J. M. Findlay. 1987. Sensitivity and criterion effects in the spatial cuing of visual attention. *Perception and Psychophysics,* 42: 383–399.

Müller, H. J., and G. W. Humphreys. 1991. Luminance-increment detection: Capacity-limited or not? *Journal of Experimental Psychology: Human Perception and Performance,* 17: 107–124.

Müller, H. J., and P. M. A. Rabbitt. 1989. Reflexive and voluntary orienting of visual attention: Time course of activation and resistance to interruption. *Journal of Experimental Psychology: General,* 15: 315–330.

Mumford, D. 1991. On the computational architecture of the neocortex I: The role of the thalamo-cortical loop. *Biological Cybernetics,* 65: 135–145.

Muñoz, D. P., and R. H. Wurtz. 1992. Two classes of cells with saccade-related activity in the monkey superior colliculus. *Society for Neuroscience Abstracts,* 18: 699.

——— 1993a. Fixation cells in monkey superior colliculus. I. Characteristics of cell discharge. *Journal of Neurophysiology,* 70: 559–575.

——— 1993b. Fixation cells in monkey superior colliculus. II. Reversible activation and deactivation. *Journal of Neurophysiology,* 70: 576–589.

Myers, G. E. 1986. *William James.* New Haven, CT: Yale University Press.

Naatanen, R. 1982. Processing negativity: An evoked-potential reflection of selective attention. *Psychological Bulletin,* 92: 605–640.

——— 1992. *Attention and Brain Function.* Hillsdale, NJ: Erlbaum.

Naatanen, R., and A. E. Merisalo. 1977. Expectancy and preparation in simple reaction time. In *Attention and Performance VI,* ed. S. Dornic, 115–138. Hillsdale, NJ: Erlbaum.

Nagel, T. 1986. *The View from Nowhere.* Oxford: Oxford University Press.

Nakamura, H., R. Gattass, R. Desimone, and L. G. Ungerleider. 1991. Comparison of inputs from areas V1 and V2 to areas V4 and TEO in macaque. *Society for Neuroscience Abstracts,* 17: 845.

———— 1992. Projections to visual areas V4 and TEO from temporal, parietal, and frontal lobes in macaques. *Society for Neuroscience Abstracts,* 19:
294.

Nakayama, K. and M. MacKeben. 1989. Sustained and transient components of focal visual attention. *Vision Research,* 29: 1631–1647.

Nakayama, K., and G. H. Silverman. 1986. Serial and parallel processing of
visual feature conjunctions. *Nature,* 320: 264–265.

Navon, D., and D. Gopher. 1980. Task difficulty, resources, and dual task
performance. In *Attention and performance VIII,* ed. R. S. Nickerson,
297–315. Hillsdale, NJ: Erlbaum.

Neill, W. T., and R. L. Westberry. 1987. Selective attention and the suppression of cognitive noise. *Journal of Experimental Psychology: Learning, Memory, and Cognition,* 13: 327–334.

Nicolelis, M. A., R. C. Lin, D. J. Woodward, and J. K. Chapin. 1993. Dynamic
and distributed properties of many-neuron ensembles in the ventral
posterior medial thalamus of awake rats. *Proceedings of the National Academy of Sciences (USA),* 90: 2212–2216.

Niemi, P., and R. Naatanen. 1981. Foreperiod and simple reaction time.
*Psychological Bulletin,* 89: 133–162.

Norman, D. A. 1968. Toward a theory of memory and attention. *Psychological
Review,* 75: 522–536.

Norman, D. A., and T. Shallice. 1980. Attention to action: Willed and automatic control of behavior. San Diego, CA: University of California, Center for Human Information Processing, Report No. 8006.

Ogasawara, K., J. G. McHaffie, and B. E. Stein. 1984. Two visual corticotectal
systems in cat. *Journal of Neurophysiology,* 52: 1226–1245.

Ogren, M. P., and A. E. Hendrickson. 1977. The distribution of pulvinar
terminals in visual areas 17 and 18 of the monkey. *Brain Research,* 137:
343–350.

Ojemann, G. 1983. Brain organization for language from the perspective
of electrical stimulation mapping. *Behavioral and Brain Sciences,* 2:
189–206.

Ollman, R. T. 1966. Fast guesses in choice-reaction time. *Psychonomic Science,*
6: 155–156.

Olshausen, B. A., C. H. Anderson, and D. C. Van Essen. 1993. A neurobiological model of visual attention and invariant pattern recognition
based on dynamic routing of information. *Journal of Neuroscience,* 13:
4700–4719.

O'Reilly, R. C., S. M. Kosslyn, C. J. Marsolek, and C. F. Chabris. 1990. Receptive field characteristics that allow parietal lobe neurons to encode
spatial properties of visual input: A computational analysis. *Journal of
Cognitive Neuroscience,* 2: 141–155.

Palmer, J. 1994. Set-size effects in visual search: The effect of attention is independent of the stimulus for simple tasks. *Vision Research*, 34: 1703–1721.

Palmer, J., C. T. Ames, and D. T. Lindsey. 1993. Measuring the effect of attention on simple visual search. *Journal of Experimental Psychology: Human Perception and Performance*, 19: 108–130.

Pandya, D. N., P. Dye, and N. Butters. 1971. Efferent cortico-cortical projections of the prefrontal cortex in the rhesus monkey. *Brain Research*, 31: 35–46.

Pandya, D. N., D. L. Rosene, and A. M. Doolittle. 1994. Corticothalamic connections of auditory-related areas of the temporal lobe in the rhesus monkey. *Journal of Comparative Neurology*, 345: 447–471.

Parasuraman, R., and D. R. Davies, eds. 1984. *Varieties of Attention*. New York: Academic Press.

Pardo, J. V., P. T. Fox, and M. E. Raichle. 1991. Localization of a human system for sustained attention by positron emission tomography. *Nature*, 349: 61–64.

Pardo, J. V., P. J. Pardo, K. W. Janer, and M. E. Raichle. 1990. The anterior cingulate cortex mediates processing selection in the Stroop attentional conflict paradigm. *Proceedings of the National Academy of Sciences (USA)*, 87: 256–259.

Pardo, J. V., G. D. Partlow, M. Colonnier, and J. Szabo. 1977. Thalamic projections of the superior colliculus in the rhesus monkey, *Macaca mulatta*. A light and electron microscopic study. *Journal of Comparative Neurology*, 171: 285–317.

Pashler, H. 1987. Detecting conjunctions of color and form: Reassessing the serial search hypothesis. *Perception and Psychophysics*, 41: 191–201.

——— 1992. Attentional limitations in doing two tasks at the same time. *Current Directions in Psychological Science*, 1: 44–48.

——— 1994. Graded capacity-sharing in dual task interference? *Journal of Experimental Psychology: Human Perception and Performance*, 20: 330–342.

Passingham, R. E. 1972. Visual discrimination learning after selective prefrontal ablations in monkeys *(Macaca mulatta)*. *Neuropsychology*, 10: 27–30.

——— 1975. Delayed matching after selective prefrontal lesions in monkeys *(Macaca mulatta)*. *Brain Research*, 92: 89–102.

Paulesu, E., C. D. Frith, and R. S. J. Frackowiak. 1993. The neural correlates of the verbal component of working memory. *Nature*, 362: 342–345.

Pavlov, I. P. 1927. *Conditioned Reflexes*. London: Clarendon Press.

Perrett, D. I., A. J. Mistlin, and A. J. Chitty. 1987. Visual cells responsive to faces. *Trends in Neurosciences*, 10: 358–364.

Perrett, D. I., A. J. Mistlin, A. J. Chitty, P. A. Smith, D. Potter, R. Broennini-mann, and M. Harries. 1985. Specialized face processing and hemisphere asymmetry in man and monkey: Evidence from single unit and reaction time studies. *Behavioural Brain Research*, 3: 245–258.

Perrett, D. I., E. T. Rolls, and W. Caan. 1982. Visual neurons responsive to faces in the monkey temporal cortex. *Experimental Brain Research*, 47: 329–342.

Perrett, D. I., P. A. J. Smith, D. D. Potter, A. J. Mistlin, A. S. Head, A. D. Milner, and M. A. Jeeves. 1985. Visual cells in the temporal cortex sensitive to face view and gaze direction. *Proceedings of the Royal Society of London* (B), 223: 293–317.

Perry, R. B. 1967. *The Thought and Character of William James*. Cambridge, MA: Harvard University Press.

Petersen, S. E., P. T. Fox, M. I. Posner, M. Minton, and M. E. Raichle. 1988. Positron emission tomographic studies of the cortical anatomy of single word processing. *Nature*, 331: 585–589.

Petersen, S. E., P. T. Fox, A. Z. Snyder, and M. E. Raichle. 1990. Activation of extrastriate and frontal cortical areas by visual words and word-like stimuli. *Science*, 249: 1041–1044.

Petersen, S. E., D. L. Robinson, and W. Keys. 1985. Pulvinar nuclei of the behaving rhesus monkey: Visual responses and their modulation. *Journal of Neurophysiology*, 54: 867–886.

Petersen, S. E., D. L. Robinson, and J. D. Morris. 1987. Contributions of the pulvinar to visual spatial attention. *Neuropsychologia*, 25: 97–105.

Petersen, S. E., P. T. Fox, A. Z. Snyder, and M. E. Raichle. 1990. Activation of extrastriate and frontal cortical areas by visual words and word-like stimuli. *Science*, 249: 1041–1044.

Phelps, M. E., and J. C. Mazziotta. 1985. Positron emission tomography: Human brain function and biochemistry. *Science*, 228: 799–809.

Pohl, W. 1973. Dissociation of spatial discrimination deficits following frontal and parietal lesions in monkeys. *Journal of Comparative and Physiological Psychology*, 82: 227–239.

Posner, M. I. 1978. *Chronometric Explorations of Mind*. Englewood Cliffs, NJ: Erlbaum.

——— 1980. Orienting of attention. *Quarterly Journal of Experimental Psychology*, 32: 3–25.

——— 1994. Attention in cognitive neuroscience: An overview. In *The Cognitive Neurosciences*, ed. M. Gazzaniga. Cambridge, MA: MIT Press.

Posner, M. I., and S. J. Boies. 1971. Components of attention. *Psychological Review*, 78: 391–408.

Posner, M. I., and Y. Cohen. 1984. Components of performance. In *Attention and Performance,* ed. H. Bouma and D. Bowhuis, 531–556. Hillsdale, NJ: Erlbaum.

Posner, M. I., A. Inhoff, F. J. Friedrich, and A. Cohen. 1987. Isolating attentional systems: A cognitive-anatomical analysis. *Psychobiology,* 15: 107–121.

Posner, M. I., M. J. Nissen, and W. C. Ogden. 1978. Attended and unattended processing modes: The role of set for spatial location. In *Modes of Perceiving and Processing Information,* ed. H. L. Pick and I. J. Saltzman. Hillsdale, NJ: Erlbaum.

Posner, M. I., and S. E. Petersen. 1990. The attention system of the human brain. *Annual Review of Neuroscience,* 13: 25–41.

Posner, M. I., S. E. Peterson, P. T. Fox, and M. E. Raichle. 1988. Localization of cognitive operations in the human brain. *Science,* 240: 1627–1631.

Posner, M. I., and D. E. Presti. 1987. Selective attention and cognitive control. *Trends in Neurosciences,* 10: 13–17.

Posner, M. I., and M. E. Raichle. 1994. *Images of Mind.* New York: Freeman.

Posner, M. I., and C. R. R. Snyder. 1975. Facilitation and inhibition in the processing of signals. In *Attention and Performance V,* ed. P. M. A. Rabbitt and S. Dornic, 669–681. New York: Academic Press.

Posner, M. I., C. R. R. Snyder, and B. J. Davidson. 1980. Attention and the detection of signals. *Journal of Experimental Psychology: General,* 109: 160–174.

Posner, M. I., J. A. Walker, F. J. Friedrich, and R. D. Rafal. 1984. Effects of parietal injury on covert orienting of attention. *Journal of Neuroscience,* 4: 1863–1874.

Possamai, C.-A., and A.-M. Bonnel. 1991. Early modulation of visual input: Constant versus varied cuing. *Bulletin of the Psychonomic Society,* 29: 323–326.

Prokasy, W. F., Jr. 1961. Nonrandom stimulus sampling in statistical learning theory. *Psychological Review,* 68: 219–224.

Pylyshyn, Z., J. Burkell, B. Fisher, C. Sears, W. Schmidt, and L. Trick. 1994. Multiple parallel access in visual attention. *Canadian Journal of Experimental Psychology,* 48: 260–282.

Rafal, R. D., P. A. Calabresi, C. W. Brennan, and T. K. Sciolto. 1989. Saccade preparation inhibits reorienting to recently attended locations. *Journal of Experimental psychology: Human Perception and Performance,* 15: 673–685.

Rafal, R. D., and M. I. Posner. 1987. Deficits in human visual spatial attention following thalamic lesions. *Proceedings of the National Academy of Sciences (USA),* 84: 7349–7353.

Rafal, R. D., M. I. Posner, J. H. Friedman, A. W. Inhoff, and E. Berstein.

1988. Orienting of visual attention in progressive supranuclear palsy. *Brain*, 111: 267–280.

Raichle, M. 1987. Circulatory and metabolic correlates of brain function in normal humans. In *Handbook of Physiology*, sec. 1, *The Nervous System*, vol. 5: *Higher Functions of the Brain, Part 2*, ed. F. Plum. Bethesda, MD: American Physiological Society.

Raichle, M. E., J. A. Fiez, T. O. Videen, A. K. MacLeod, J. V. Pardo, P. T. Fox, and S. E. Petersen. (in press). Practice-related changes in human brain functional anatomy during non-motor learning. *Cerebral Cortex*, 4.

Raichle, M. E., R. L. Grubb, M. H. Gado, J. O. Eichling, and M. M. Ter-Pogossian. 1976. Correlation between regional cerebral blood flow and oxidative metabolism. *Archives of Neurology*, 33: 523–526.

Redies, H., S. Brandner, and O. D. Creutzfeld. 1989. Anatomy of the auditory thalamocortical system of the guinea pig. *Journal of Comparative Neurology*, 282: 489–511.

Reeves A., and G. Sperling. 1986. Attention gating in short-term visual memory. *Psychological Review*, 93: 180–206.

Remington, R., and L. Pierce. 1984. Moving attention: Evidence for time invariant shifts of visual selective attention. *Perception and Psychophysics*, 35: 393–399.

Rhoades, R. W., R. D. Mooney, W. H. Rohrer, N. M. Nikoletseas, and S. E. Fish. 1989. Organization of the projection from the superficial to the deep layers of the hamster's superior colliculus as demonstrated by the anterograde transport of *Phaseolus vulgaris* leucoagglutinin. *Journal of Comparative Neurology*, 283: 54–70.

Richmond, B J., and T. Sato. 1987. Enhancement of inferior temporal neurons during visual discrimination. *Journal of Neurophysiology*, 58: 1292–1306.

Rizzolatti, G., M. Gentilucci, and M. Matelli. 1985. Selective spatial attention: One center, one circuit or many circuits? In *Attention and Performance XI*, ed. M. I. Posner and O. S. M. Marin, 251–265. Hillsdale, NY: Erlbaum.

Rizzolatti, G., L. Riggio, I. Dascola, and C. Umilta. 1987. Reorienting attention across the horizontal and vertical meridians: Evidence in favor of a premotor theory of attention. *Neuropsychologia*, 25 (1A): 31–40.

Robbins, T W. and B. J. Everitt. 1994. Arousal systems and attention. In *The Cognitive Neurosciences*, ed M. S. Gazzaniga. Cambridge, MA: MIT Press.

Robertson, L., M. R. Lamb, and R. T. Knight. 1988. Effects of lesions of temporal-parietal junction on perceptual and attentional processing in humans. *Journal of Neuroscience*, 8: 3757–3769.

Robinson, D. A. 1970. Oculomotor unit behavior in the monkey. *Journal of Neurophysiology*, 33: 393–404.

———— 1992. Implications of neural networks for how we think about brain function. *Behavioral and Brain Sciences*, 15: 644–655.

Robinson, D. L., M. E. Goldberg, and G. B. Stanton. 1978. Parietal association cortex in the primate: Sensory mechanisms and behavioral modulations. *Journal of Neurophysiology*, 41: 910–932.

Robinson, D. L., and S. E. Petersen. 1992. The pulvinar and visual salience. *Trends in Neurosciences*, 15: 127–132.

Robinson, D. L., S. E. Petersen, and W. Keys. 1986. Sacccade-related and visual activites in the pulvinar nuclei of the behaving rhesus monkey. *Experimental Brain Research*, 62: 625–634.

Robinson, D. L., and M. D. Rugg. 1988. Latencies of visually responsive neurons in various regions of the rhesus monkey brain and their relation to human visual responses. *Biological Psychology*, 26: 111–116.

Rockland, K. S., and D. N. Pandya. 1979. Laminar origins and terminations of cortical connections of the occipital lobe in the rhesus monkey. *Brain Research*, 179: 3–20.

Rohrbaugh, J. W., K. Syndulko, and D. B. Lindsley. 1976. Brain wave components of the contingent negative variation in humans. *Science*, 191: 1055–1057.

Roland, P. E. 1985. Cortical organization of voluntary behavior in man. *Human Neurobiology*, 4: 155–167.

———— 1993. *Brain Activation*. New York: Wiley-Liss.

Rosch, E. 1976. Classifications of objects in the real world: Origins and representations in cognition. *Bulletin de Psychologie*, special annual volume, 242–250.

Rosenkilde, C. E., R. H. Bauer, and J. M. Fuster. 1981. Single cell activity in ventral prefrontal cortex of behaving monkeys. *Brain Research*, 209: 275–294.

Rueckl, J. G., K. R. Cave, and S. M. Kosslyn. 1989. Why are "what" and "where" processed by separate cortical visual systems? A computational investigation. *Journal of Cognitive Neuroscience*, 1: 171–186.

Rumelhart, D. E., and J. L. McClelland, eds. 1986. *Parallel Distributed Processing: Explorations in the Microstructure of Cognition*, vol. 1. Cambridge, MA: MIT Press/Bradford Books.

Saarinen, J., and B. Julesz. 1991. The speed of attentional shifts in the visual field. *Proceedings of the National Academy of Sciences (USA)*, 88: 1812–1814.

Sandon, P. A. 1990. Simulating visual attention. *Journal of Cognitive Neuroscience*, 2: 213–231.

Scheibel, A. B. 1981. The problem of selective attention: A possible structural substrate. In *Brain Mechanisms and Perceptual Awareness,* ed. O. Pompeiano and C. Marsen. New York: Raven Press.

Schiller, P. H. 1970. The discharge characteristics of single units in the oculomotor and abducens nuclei of the unanesthetized monkey. *Experimental Brain Research,* 10: 347–362.

———— 1984. The superior colliculus and visual function. In *Handbook of Physiology,* sec. 1, *The Nervous System,* vol. 3: *Sensory Processes,* ed. J. M. Brookhart and V. B. Mountcastle, 457–505. Bethesda, MD: American Physiological Society.

———— 1985. A model for the generation of visually guided saccadic eye movements. In *Models of the Visual Cortex,* ed. D. Rose and V. G. Dobson. Chichester: Wiley.

Schiller, P. H., and F. Koerner. 1971. Discharge characteristics of single units in the superior colliculus of the alert rhesus monkey. *Journal of Neurophysiology,* 34: 920–936.

Schiller, P. H., and K. Lee. 1991. The role of the primate extrastriate area V4 in vision. *Science,* 251: 1251–1253.

Schiller, P. H., N. K. Logothetis, and E. R. Charles. 1990. Functions of the colour-opponent and broad-band channels of the visual system. *Nature,* 343: 68–70.

Schiller, P. H., and J. Malpeli. 1977. Properties and tectal projections of monkey retinal ganglion cells. *Journal of Neurophysiology,* 40: 428–445.

Schiller, P. H., J. H. Sandell, and J. H. Maunsell. 1987. The effect of frontal eye field and superior colliculus lesions on saccadic latencies in the rhesus monkey. *Journal of Neurophysiology,* 57: 1033–1049.

Schiller, P. H., and M. Stryker. 1972. Single-unit recording and stimulation in superior colliculus of the alert rhesus monkey. *Journal of Neurophysiology,* 35: 915–924.

Schmahmann, J. D., and D. N. Pandya. 1990. Anatomical investigation of projections from thalamus to posterior parietal cortex in the rhesus monkey: A WGA-HRP and fluorescent tracer study. *Journal of Comparative Neurology,* 295: 299–326.

Schneider, W., and R. M. Shiffrin. 1977. Controlled and automatic human information processing: I. Detection, search, and attention. *Psychological Review,* 84: 1–66.

Schwartz, M. L., J. J. Dekker, and P. S. Goldman-Rakic. 1991. Dual mode of corticothalamic synaptic termination in the mediodorsal nucleus of the rhesus monkey. *Journal of Comparative Neurology,* 309: 289–304.

Schwartz, E. L., R. Desimone, T. D. Albright, and C. G. Gross. 1983. Shape recognition and inferior temporal neurons. *Proceedings of the National Academy of Sciences (USA),* 80: 5776–5778.

Searle, J. 1980. Minds, brains, and programs. *Behavioral and Brain Sciences,* 3: 417–458.

Selemon, L. D., and P. S. Goldman-Rakic. 1988. Common cortical and sub-cortical target areas of the dorsolateral prefrontal and posterior parietal cortices in the rhesus monkey: Evidence for a distributed neural network subserving spatially guided behavior. *Journal of Neuroscience,* 8: 4049–4068.

Seltzer, B., and D. N. Pandya. 1980. Afferent cortical connections and architectonics of the superior temporal sulcus and surrounding cortex in the rhesus monkey. *Brain Research,* 149: 1–24.

——— 1984. Further observations on parieto-temporal connections in the rhesus monkey. *Experimental Brain Research,* 55: 301–312.

Shallice, T. 1988. *From Neuropsychology to Mental Structure.* Cambridge: Cambridge University Press.

Shannon, C. E., and W. Weaver. 1949. *The Mathematical Theory of Communication.* Urbana: University of Illinois Press.

Shaw, M. L. 1978. A capacity allocation model for reaction time. *Journal of Experimental Psychology: Human Perception and Performance,* 4: 596–598.

——— 1984. Division of attention among spatial locations: A fundamental difference between detection of letters and detection of luminance increments. In *Attention and Performance X,* ed. H. Bouma and D. G. Bouwhuis. Hillsdale, NJ: Erlbaum.

Shaw, M. L., and P. Shaw. 1978. A capacity allocation model for reaction time. *Journal of Experimental Psychology: Human Perception and Performance,* 4: 586–598.

Sheliga, B. M., L. Riggio, and G. Rizzolatti. 1994. Orienting of attention and eye movements. *Experimental Brain Research,* 98: 507–522.

Shepard, R. N. 1987. Toward a universal law of generalization for psychological science. *Science,* 237: 1317–1323.

Shepherd, G. M. 1990. *The Synaptic Organization of the Brain.* New York: Oxford University Press.

Sherman, S. M., and C. Koch. 1986. The control of retinogeniculate transmission in the mammalian lateral geniculate nucleus. *Experimental Brain Research,* 63: 1–20.

——— 1990. Thalamus. In *The Synaptic Organization of the Brain,* ed. G. M. Shepherd. New York: Oxford University Press.

Shiffrin, R. M. 1988. Attention. In *Steven's Handbook of Experimental Psychology,* 2d ed., ed. R. C. Atkinson, R. J. Herrnstein, G. Lindsey, and R. D. Luce. New York: Wiley.

Shiffrin, R. M., and W. Schneider. 1977. Controlled and automatic human information processing. II. Perceptual learning, automatic attending, and a general theory. *Psychological Review,* 84: 127–190.

Shulman, G. L., R. W. Remington, and J. P. McLean. 1979. Moving attention through visual space. *Journal of Experimental Psychology*, 5: 522–526.

Sieroff, E., and M. I. Posner. 1988. Cueing spatial attention during processing of words and letter strings in normals. *Cognitive Neuropsychology*, 5: 451–472.

Singer, W. 1977. Control of thalamic transmission by corticofugal and ascending reticular pathways in the visual system. *Physiological Reviews*, 57: 386–420.

——— 1990. Search for coherence: A basic principle of cortical self-organization. *Concepts in Neuroscience*, 1: 1–26.

Sparks, D. L. 1986. Translation of sensory signals into commands for control of saccadic eye movements: Role of primate superior colliculus. *Physiological Reviews*, 66: 118–171.

Sparks, D. L., R. Holland, and B. L. Guthrie. 1976. Size and distribution of movement fields in the monkey superior colliculus. *Brain Research*, 113: 21–34.

Sparks, D., and L. Mays. 1980. Movement fields of saccade-related burst neurons in the monkey superior colliculus. *Brain Research*, 190: 39–50.

Sparks, D. L, and J. S. Nelson. 1987. Sensory and motor maps in the mammalian superior colliculus. *Trends in Neurosciences*, 10: 312–317.

Sparks, D. L., and J. G. Pollack. 1977. The neural control of saccadic eye movements: The role of the superior colliculus. In *Eye Movements*, ed. B. A. Brooks and F. J. Bajandas, 179–219. New York: Plenum.

Sperling, G. 1984. A unified theory of attention and signal detection. In *Varieties of Attention*, ed. R. Parasuraman and D. R. Davies, 103–181. New York: Academic Press.

Sperling, G., and B. A. Dosher. 1986. Strategy and optimization in human information processing. In *Handbook of Perception and Performance*, vol. 1, ed. K. Boff, L. Kaufman, and J. Thomas, 2.1–2.65. New York: Wiley.

Spiegler, B. J., and M. Mishkin. 1981. Evidence for the sequential participation of inferior temporal cortex and amygdala in the acquisition of stimulus-reward associations. *Behavioral Brain Research*, 3: 303–317.

Spitzer, H., R. Desimone, and J. Moran. 1988. Increased attention enhances both behavioral and neuronal performance. *Science*, 240: 338–340.

Stanton, G. B, M. E. Goldberg, and C. J. Bruce. 1988. Frontal eye fields in the macaque monkey: I. Subcortical pathways and topography of striatal and thalamic terminal fields. *Journal of Comparative Neurology*, 271: 473–492.

Stein, B. E., and M. O. Arigbede. 1972. Unimodal and multimodal response properties of neurons in the cat's superior colliculus. *Experimental Neurology*, 36: 179–196.

Stein, B. E., B. Magalhaes-Castro, and I. Kruger. 1976. Relationship between visual and tactile representation in cat superior colliculus. *Journal of Neurophysiology*, 39: 401–419.

Stein, J. F. 1978. Effects of parietal lobe cooling on manipulation in the monkey. In *Active Touch*, ed. G. Gordon. Oxford: Pergamon Press.

———— 1992. The representation of egocentric space in the posterior parietal cortex. *Behavioral and Brain Sciences*, 15: 691–700.

Steriade, M., D. Contreras, R. Curró Dossi, and A. Nuñez. 1993. The slow (<1 Hz) oscillation in reticular thalamic and thalamocortical neurons: Scenario of sleep rhythm generation in interacting thalamic and neocortical networks. *Journal of Neuroscience*, 13: 3284–3299.

Steriade, M., L. Domich, and G. Oakson. 1986. Reticularis thalami neurons revisited: Activity changes during shifts in states of vigilance. *Journal of Neuroscience*, 6: 68–81.

Steriade, M., E. G. Jones, and R. R. Llinas. 1990. *Thalamic Oscillations and Signaling*. New York: Wiley.

Steriade, M., D. A. McCormick, and T. J. Sejnowski. 1993. Thalamocortical oscillations in the sleeping and aroused brain. *Science*, 262: 679–685.

Sternberg, S. 1966. High-speed scanning in human memory. *Science*, 153: 652–654.

Swensen, R. G., and W. Edwards. 1971. Response strategies in a two-choice reaction task with a continuous cost for time. *Journal of Experimental Psychology*, 88: 67–81.

Takacs, J., J. Hamori, and V. Silakov. 1991. GABA-containing neuronal processes in normal and cortically deafferented dorsal lateral geniculate nucleus of the cat: An immunogold and quantitative EM study. *Experimental Brain Research*, 83: 562–574.

Tanaka, K., H. Saito, Y. Fukada, and M. Moriya. 1991. Coding visual images of objects in the inferotemporal cortex of the macaque monkey. *Journal of Neurophysiology*, 66: 170–189.

Tanner, W. P., Jr., and J. A. Swets. 1954. A decision-making theory of visual detection. *Psychological Review*, 61: 401–409.

Tassinari, G., S. Aglioti, L. Chelazzi, C. A. Marzi, and G. Berlucci. 1987. Distribution in the visual field of the costs of voluntarily allocated attention and of the inhibitory after-effects of covert orienting. *Neuropsychologia*, 25: 55–72.

Theeuwes, J. 1991. Exogenous and endogenous control of attention: The effect of visual onsets and offsets. *Perception and Psychophysics*, 49: 83–90.

———— 1992. Perceptual selectivity for color and form. *Perception and Psychophysics*, 51: 599–606.

Tipper, S. P. 1985. The negative priming effect: Inhibitory priming by ignored objects. *Quarterly Journal of Experimental Psychology*, 37A: 571–590.

Tovée, M. J., E. T. Ralls, A. Treves, and R. P. Bellis. 1993. Information encoding and the responses of single neurons in the primate temporal visual cortex. *Journal of Neurophysiology,* 70: 640–654.

Townsend, J. T. 1976. Serial and within-stage independent parallel model equivalence on the minimal completion time. *Journal of Mathematical Psychology,* 14: 219–239.

Toyama, K. 1988. Functional connections of the visual cortex studied by cross-correlation techniques. In *Neurobiology of Neocortex,* ed. P. Rakic and W. Singer, 203–217. New York: Wiley.

Treisman, A. 1960. Contextual cues in selective listening. *Quarterly Journal of Experimental Psychology,* 12: 242–248.

Treisman, A. M. 1964a. Monitoring and storage of irrelevant messages in selective attention. *Journal of Verbal Learning and Verbal Behavior,* 3: 449–459.

———— 1964b. Verbal cues, language, and meaning in selective attention. *American Journal of Psychology,* 77: 206–219.

Treisman, A. 1985. Preattentive processing in vision. *Computer Vision, Graphics, and Image Processing,* 31: 156–177.

———— 1988. Features and objects: The fourteenth Bartlett memorial lecture. *Quarterly Journal of Experimental Psychology,* 40A: 201–237.

Treisman, A., and G. Gelade. 1980. A feature integration theory of attention. *Cognitive Psychology,* 12: 97–136.

Treisman, A., and S. Gormican. 1988. Feature analysis in early vision: Evidence from search asymmetries. *Psychological Review,* 95: 15–48.

Treisman, A., and S. Sato. 1990. Conjunction search revisited. *Journal of Experimental Psychology: Human Perception and Performance,* 16: 459–478.

Trick, L. M., and Z. W. Pylyshyn. 1993. What enumeration studies can show us about spatial attention: Evidence for limited capacity preattentive processing. *Journal of Experimental Psychology: Human Perception and Performance,* 19: 331–351.

Trojanowski, J. Q., and S. Jacobsen. 1977. The morphology and laminar distribution of cortico-pulvinar neurons in the rhesus monkey. *Experimental Brain Research,* 28: 51–62.

Tsal, Y. 1983. Movement of attention across the visual field *Journal of Experimental Psychology: Human Perception and Performance,* 9: 523–530.

Ungerleider, L. G., and C. A. Christensen. 1979. Pulvinar lesions in monkeys produce abnormal scanning of a complex visual array. *Neuropsychologia,* 17: 493–501.

Ungerleider, L. G., D. Gaffan, and V. S. Pelak. 1989. Projections from inferior temporal cortex to prefrontal cortex via the uncinate fascicle in rhesus monkeys. *Experimental Brain Research,* 76: 473–484.

Ungerleider, L. G., T. W. Galkin, and M. Mishkin. 1983. Visuotopic organization of projections from striate cortex to inferior and lateral pulvinar in rhesus monkey. *Journal of Comparative Neurology*, 217: 137–157.

Ungerleider, L. G., R. Gattass, A. P. B. Sousa, and M. Mishkin. 1983. Projections of area V2 in the macaque. *Society for Neuroscience Abstracts*, 9: 152.

Ungerleider, L. G., and M. Mishkin. 1982. Two cortical visual systems. In *Analysis of Visual Behavior*, ed. D. J. Ingle, M. A. Goodale, and R. J. W. Mansfield, 549–586. Cambridge, MA: MIT Press.

Unyk, A. M., and J. C. Carlsen. 1987. The influence of expectancy on melodic perception. *Psychomusicology*, 7: 3–23.

Van Buren, J. M., and R. C. Borke. 1972. *Varieties and Connections of the Human Thalamus*. New York: Springer-Verlag.

Van der Heijden, A. H. C. 1992. *Selective Attention in Vision*. London: Routledge.

Van der Heijden, A. H. C., and E. Eerland. 1973. The effects of cueing in a visual signal detection task. *Quarterly Journal of Experimental Psychology*, 25: 496–503.

Van Essen, D. C. 1985. Functional organization of primate visual cortex. In *Cerebral Cortex*, vol. 3, ed. A. Peters and E.G. Jones, 259–329. New York: Plenum.

Van Essen, D. C., C. H. Anderson, and D. J. Felleman. 1992. Information processing in the primate visual system: An integrated systems perspective. *Science*, 255: 419–423.

Van Essen, D. C., and J. H. R. Maunsell. 1980. Two-dimensional maps of the cerebral cortex. *Journal of Comparative Neurology*, 191: 255–281.

Vecera, S. P. and M. J. Farah. 1994. Does visual attention select objects or locations? *Journal of Experimental Psychology: General*, 123, 146–160.

Von Bonin, G., and P. Bailey. 1947. *The Neocortex of Macaca mulatta*. Urbana: University of Illinois Press.

Wallace, M. T., M. A. Meredith, and B. E. Stein. 1991. Cortical convergence of multisensory output neurons of cat superior colliculus. *Society for Neuroscience Abstracts*, 17: 1379.

Walley, R. E., and T. D. Weiden. 1973. Lateral inhibition and cognitive masking: A neuropsychological theory of attention. *Psychological Review*, 80: 284–302.

Wang, Z., and D. A. McCormick. 1993. Control of firing mode of corticotectal and corticopontine layer V burst-generating neurons by norepinephrine. *Journal of Neuroscience*, 13: 2199–2216.

Watson, C. 1985. *Basic Human Neuroanatomy*. Boston: Little, Brown and Company.

Weber, J. T., and T. C. T. Yin. 1984. Subcortical projections of the inferior parietal cortex (area 7) in the stump-tailed monkey. *Journal of Comparative Neurology*, 224: 206–230.

Webster, M. J., L. G. Ungerleider, and J. Bachevalier. 1991. Subcortical connections of inferior temporal areas TE and TEO in macaques. *Society for Neuroscience Abstracts*, 17: 845.

Weichselgartner, E., and G. Sperling. 1987. Dynamics of automatic and controlled visual attention. *Science*, 238: 778–780.

Welford, A. T. 1952. The "psychological refractory period" and the timing of high speed performance—A review and a theory. *British Journal of Psychology*, 43: 2–19.

——— 1967. Single-channel operation in the brain. *Acta Psychologica*, 27: 5–22.

Wickens, C. D. 1984. Processing resources in attention. In *Varieties of Attention*, ed. R. Parasuraman and D. R. Davies, 63–102. Orlando, FL: Academic Press.

Wilson, F. A. W., S. P. O'Scalaidhe, and P. S. Goldman-Rakic. 1993. Dissociation of object and spatial processing domains in primate prefrontal cortex. *Science*, 260: 1955–1958.

Woldorff, M. G., C. C. Gallen, S. A. Hampson, S. A. Hillyard, C. Pantev, D. Sobel, and F. E. Bloom. 1993. Modulation of early sensory processing in human auditory cortex during auditory selective attention. *Proceedings of the National Academy of Sciences (USA)*, 90: 8722–8726.

Wolfe, J., K. R. Cave, and S. L. Franzel. 1989. Guided search: An alternative to the feature integration model for visual search. *Journal of Experimental Psychology: Human Perception and Performance*, 419–433.

Wright, R. D. 1994. Shifts of visual attention to multiple simultaneous location cues. *Canadian Journal of Experimental Psychology*, 48: 205–217.

Wright, R. D., and L. M. Ward. 1994. Indexing and the control of express saccades. *Behavioral and Brain Sciences*, 16: 594–595.

Wurtz, R. H. 1969. Visual receptive fields of striate cortex neurons in awake monkeys. *Journal of Neurophysiology*, 32: 727–742.

Wurtz, R. H., and M. E. Goldberg. 1972a. Activity of superior colliculus in behaving monkey. III. Cells discharging before eye movements. *Journal of Neurophysiology*, 35: 575–586.

——— 1972b. Activity of superior colliculus in behaving monkey. IV. Effects of lesions on eye movements. *Journal of Neurophysiology*, 35: 587–596.

Wurtz, R. H., M. E. Goldberg, and D. L. Robinson. 1980. Behavioral modulation of visual responses in the monkey: Stimulus selection for attention and movement. *Progress in Psychobiology and Physiological Psychology*, 9: 43–83.

Wurtz, R. H., and C. W. Mohler. 1976. Organization of monkey superior colliculus: Enhanced visual responses of superficial layer cells. *Journal of Neurophysiology*, 39: 745–762.

Yajeya, J., J. Quintana, and J. M. Fuster. 1988. Prefrontal representation of stimulus attributes during delay tasks. II. The role of behavior significance. *Brain Research*, 474: 222.

Yantis, S. 1993a. Stimulus-driven attentional capture. *Current Directions in Psychological Science*, 2: 156.

—— 1993b. Stimulus-driven attentional capture and attentional control settings. *Journal of Experimental Psychology: Human Perception and Performance*, 19: 676–681.

Yantis, S., and J. Jonides. 1990. Abrupt visual onsets and selective attention: Voluntary versus automatic allocation. *journal of Experimental Psychology: Human Perception and Performance*, 16: 121–134.

Yeh, Y., and C. W. Eriksen. 1984. Name codes and features in the discrimination of letter forms. *Perception and Psychophysics*, 36: 225–233.

Yellott, J. I., Jr. 1971. Correction for guessing and the speed-accuracy tradeoff in choice reaction time. *Journal of Mathematical Psychology*, 8: 159–199.

Yen, C. G., M. Conley, S. H. C. Hendry, and E. G. Jones. 1985. The morphology of physiologically identified GABAergic neurons in the somatic sensory part of the thalamic reticular nucleus in the cat. *Journal of Neuroscience*, 5: 2254–2268.

Yeterian, E. H., and D. N. Pandya. 1985. Corticothalamic connections of the posterior parietal cortex in the rhesus monkey. *Journal of Comparative Neurology*, 237: 408–426.

Yin, T. C. T., and V. B. Mountcastle. 1977. Visual input to the visuomotor mechanisms of the monkey's parietal lobe. *Science*, 197: 1381.

Yingling, C. D., and J. E. Skinner. 1977. Gating of thalamic input to cerebral cortex by nucleus reticularis thalami. In *Attention, Voluntary Contraction and Event-related Cerebral Potentials: Progress in Clinical Neurophysiology* (BF 1), ed. F. F. Desmedt. Basel: Karper.

Young, M. P., and S. Yamane. 1992. Sparse population coding of faces in the inferotemporal cortex. *Science*, 256: 1327–1331.

Zeki, S. M. 1969. Representation of central visual fields in prestriate cortex of monkey. *Brain Research*, 14: 271–291.

—— 1971. Cortical projections from two striate areas in the monkey. *Brain Research*, 34: 19–35.

—— 1975. The functional organization of projections from striate to prestriate visual cortex in the rhesus monkey. *Cold Spring Harbor Symposium on Quantitative Biology*, 40: 591–600.

Zeki, S., and S. Shipp. 1988. The functional logic of cortical connections. *Nature*, 335: 311–317.

—— 1989. Modular connections between areas V2 and V4 of macaque monkey visual cortex. *European Journal of Neuroscience*, 1: 494–506.

Zeki, S., D. Von Cramon, and N. Mai. 1983. Selective disturbance of move-
    ment vision after bilateral brain damage. *Brain,* 106: 313–340.
Zipser, D., and R. A. Andersen. 1988. A back-propagation programmed net-
    work that simulates response properties of a subset of posterior parietal
    neurons. *Nature,* 331: 697–684.

# *Index*

Actions, 9, 45–46, 201; executive, 137; organizational anchoring of, 9, 45–46, 203; organizational processing of, 137; planning, 201, 212

Activity distribution, 3, 66–69, 90, 203, 206–207

Algorithm(s), 5, 14–17, 39, 120, 192, 197. *See also* Attention, algorithm of

Anterior cingulate area, 108, 134–135, 168, 187

Attention: activity distribution, *see* Activity distribution; algorithm of, 16; as allocation of resources, 33, 89; as a "bottleneck," 34; as "effort," 3; as emphasis, 2, 199–202; area of, 30–33; attenuation of, 158; attributes of, 116–117, 170–171, 212; auditory, 118–119; control of, 136–139, 210–212; decrement of, 120; distributed circuits of, 212; divided, 90; early selection, 24; enhancement of, 117, 150, 158, 192, 195, 198, 210–211; expression of, 4–5, 14–16, 39, 107, 136–139, 185, 192–193, 210–212; fluctuation of, 30, 37; goals of, 8–12, *see also* Goals; intensity of, 33–34, 185–186, 207; interruption of, 65–66; late selection, 24, 43; manifestations of, 12–14; mechanisms of, 210–212; metaphors for, 3, 25–26, 38–39, 90; modulation of, 123, 138, 159; movement of, *see* Attention, shifts of; plasticity of, 214; potentia-

tion function of, 219; prolongation function of, 218–219; protective function of, 9, 137, 218, 219; resolution of, 155–156; shapes, 115–122; shifts of, 63–64, 69–87, 90, 97, 141–142, 152, 166–167; space, 116, 123, 132–136; suppression of, 153, 219; sustained, 11, 35–38, 65; synonyms for, 95; target/distractor spacing, 28, 120–122, 197; transient, 65, 211; voluntary, 131, 166; V1-to-IT pathway, 107, 120, 153, 158, 193, 209–210; V4, 115–117, 120–122, 153

Attribute processing, 101, 104, 133, 136, 156, 166, 170–171

Auditory areas, 119

Automatic processing, 1, 12, 44, 96, 99

Basal ganglia, 140, 144–145, 151, 153, 163, 165–166, 174, 196

Behaviorism, 21–22

Binding features, 16, 178. *See also* Conjoining features

Brain imaging, 13, 50, 134–136, 138, 160, 167–171, 198

Burst-firing, 178, 180–186

Channel model of attention, 3, 23, 34

Coding, 7, 128–129; attribute, *see* Attribute processing; coarse, *see* Coding, distributed; distributed, 100, 111, 127–129, 131, 145, 205; local, 88,

259

# DATE DUE

| | | | |
|---|---|---|---|
| | | | |
| | | | |
| | | | |
| | | | |
| | | | |
| | | | |
| | | | |
| | | | |
| | | | |
| | | | |
| | | | |
| | | | |
| | | | |
| | | | |
| | | | |
| | | | |
| | | | |
| | | | |
| GAYLORD | | | PRINTED IN U.S.A. |